T0205235

METHODS IN MOLECULAR BIOLOGY

Series Editor
John M. Walker
School of Life and Medical Sciences
University of Hertfordshire
Hatfield, Hertfordshire, UK

For further volumes:
http://www.springer.com/series/7651

For over 35 years, biological scientists have come to rely on the research protocols and methodologies in the critically acclaimed *Methods in Molecular Biology* series. The series was the first to introduce the step-by-step protocols approach that has become the standard in all biomedical protocol publishing. Each protocol is provided in readily-reproducible step-by-step fashion, opening with an introductory overview, a list of the materials and reagents needed to complete the experiment, and followed by a detailed procedure that is supported with a helpful notes section offering tips and tricks of the trade as well as troubleshooting advice. These hallmark features were introduced by series editor Dr. John Walker and constitute the key ingredient in each and every volume of the *Methods in Molecular Biology* series. Tested and trusted, comprehensive and reliable, all protocols from the series are indexed in PubMed.

Computer-Aided Tissue Engineering

Methods and Protocols

Edited by

Alberto Rainer

Università Campus Bio-Medico di Roma, Rome, Italy

Lorenzo Moroni

Department of Complex Tissue Regeneration, MERLN Institute for Technology Inspired Regenerative Medicine, Maastricht, The Netherlands

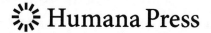 Humana Press

Editors
Alberto Rainer
Università Campus Bio-Medico
di Roma
Rome, Italy

Lorenzo Moroni
Department of Complex Tissue Regeneration
MERLN Institute for Technology Inspired
Regenerative Medicine
Maastricht, The Netherlands

ISSN 1064-3745   ISSN 1940-6029   (electronic)
Methods in Molecular Biology
ISBN 978-1-0716-0613-1   ISBN 978-1-0716-0611-7   (eBook)
https://doi.org/10.1007/978-1-0716-0611-7

This Humana imprint is published by the registered company Springer Science+Business Media, LLC, part of Springer Nature.
The registered company address is: 1 New York Plaza, New York, NY 10004, U.S.A.

Preface

With the advent of additive manufacturing technologies in the tissue engineering field, the design and fabrication of 3D biological constructs have been increasingly automated exploiting computer-aided design and manufacturing principles. This has allowed the generation of more rationally designed biofabricated products, either as additive manufactured scaffolds with instructive properties able to steer cell activity or as bioprinted constructs. In doing so, several technologies have been used and further developed, and new biomaterials synthesized, to account for the specific requirements in tissue engineering and regenerative medicine.

In this book, we have assembled a series of protocols that encompass different aspects of computer-aided design and manufacturing of 3D scaffolds and biofabricated constructs for tissue engineering applications.

The book has been divided into four sections. Section I deals with design principles. The chapters by Almeida and Bártolo, Blanquer and Grijpma, and Bonfanti et al. provide insights on topological optimization of scaffold architectures for computer-aided tissue engineering. Section II shifts the focus on synthetic routes to biomaterials compatible with additive manufacturing. Costantini et al. provide a comprehensive overview on the functionalization of naturally derived biopolymers for the formulation of photo-crosslinkable bioinks. Ronca et al. introduce a protocol for the synthesis of a polycaprolactone derivative for stereolithography. The chapter authored by Cidonio et al. discloses a protocol for the formulation of nanoclay-based inks for bone tissue engineering. Section III focuses on technological platforms and manufacturing processes. The chapter by Calore et al. provides a thorough insight on extrusion-based methods for the processing of thermoplasts. The chapters authored by Puppi and Chiellini, Li et al., and Bolle et al. introduce variants of electro-hydro-dynamic additive manufacturing techniques. Giannitelli et al. and Ghanizadeh Tabriz et al. present two different extrusion-based methods for bioprinting of cell-laden hydrogels. Last, Guzzi et al. disclose the surface tension-assisted fabrication of multicomponent structures. As a conclusion, Section IV presents relevant applicative scenarios, as in the case of the inclusion of vasculature in additively manufactured constructs, reported by Zhu et al., or in the examples of computer-aided tissue engineering for in vitro toxicology, presented by Datta et al.

We hope that you will enjoy reading this collection.

Rome, Italy   *Alberto Rainer*
Maastricht, Limburg, The Netherlands   *Lorenzo Moroni*

Contents

Contributors

HENRIQUE A. ALMEIDA • *Research Center for Information Technology and Communications, School of Technology and Management, Polytechnic Institute of Leiria, Leiria, Portugal*

LUIGI AMBROSIO • *Institute of Polymers, Composites and Biomaterials - National Research Council (IPCB-CNR), Naples, Italy*

ANDREA BARBETTA • *Department of Chemistry, Sapienza University of Rome, Rome, Italy*

PAULO J. BÁRTOLO • *Mechanical and Aeronautical Engineering Division, School of Mechanical, Aerospace & Civil Engineering, Manchester Institute of Biotechnology, Faculty of Science and Engineering, University of Manchester, Manchester, UK*

KATRIEN V. BERNAERTS • *Aachen-Maastricht Institute for Biobased Materials (AMIBM), Maastricht University, Brightlands Chemelot Campus, Geleen, The Netherlands*

ATUL BHASKAR • *Faculty of Engineering and the Environment, Southampton Innovation Boldrewood Campus, University of Southampton, Southampton, UK*

SEBASTIEN B. G. BLANQUER • *ICGM, University of Montpellier, CNRS, ENSCM, Montpellier, France*

ELEONORE C. L. BOLLE • *Institute of Health and Biomedical Innovation, Queensland University of Technology, Brisbane, QLD, Australia*

ALESSANDRA BONFANTI • *Department of Engineering, University of Cambridge, Cambridge, UK*

ANDREA ROBERTO CALORE • *Department of Complex Tissue Regeneration, MERLN Institute for Technology Inspired Regenerative Medicine, Maastricht, The Netherlands; Aachen-Maastricht Institute for Biobased Materials (AMIBM), Maastricht University, Brightlands Chemelot Campus, Geleen, The Netherlands*

SHAOCHEN CHEN • *Department of NanoEngineering, University of California, San Diego (UCSD), La Jolla, CA, USA*

FEDERICA CHIELLINI • *BIOLab Research Group, Department of Chemistry and Industrial Chemistry, University of Pisa, UdR INSTM Pisa, Pisa, Italy*

VALERIA CHIONO • *Department of Mechanical and Aerospace Engineering, Politecnico di Torino, Turin, Italy*

GIANLUCA CIDONIO • *Bone and Joint Research Group, Centre for Human Development, Stem Cells and Regeneration, Institute of Developmental Sciences, University of Southampton, Southampton, UK*

DIRK-JAN CORNELISSEN • *Institute of Mechanical Process and Energy Engineering, Heriot-Watt University, Edinburgh, UK; Department of Biomedical Engineering, University of Strathclyde, Glasgow, UK*

MARCO COSTANTINI • *Institute of Physical Chemistry, Polish Academy of Sciences, Warsaw, Poland*

PAUL D. DALTON • *Department for Functional Materials in Medicine and Dentistry and Bavarian Polymer Institute, University of Würzburg, Würzburg, Germany*

TIM R. DARGAVILLE • *Institute of Health and Biomedical Innovation, Queensland University of Technology, Brisbane, QLD, Australia*

PALLAB DATTA • *Centre for Healthcare Science and Technology, Indian Institute of Engineering Science and Technology Shibpur, Howrah, West Bengal, India*

JONATHAN I. DAWSON • *Bone and Joint Research Group, Centre for Human Development, Stem Cells and Regeneration, Institute of Developmental Sciences, University of Southampton, Southampton, UK*

LORIS DOMENICALE • *Faculty of Engineering and the Environment, Southampton Innovation Boldrewood Campus, University of Southampton, Southampton, UK*

GIUSEPPE FORTE • *Department of Materials, Holywell Park, Loughborough University, Leicestershire, UK*

CESARE GARGIOLI • *University of Rome "Tor Vergata", Rome, Italy*

SARA MARIA GIANNITELLI • *Università Campus Bio-Medico di Roma, Rome, Italy*

MICHAEL GLINKA • *Bone and Joint Research Group, Centre for Human Development, Stem Cells and Regeneration, Institute of Developmental Sciences, University of Southampton, Southampton, UK*

DIRK W. GRIJPMA • *MIRA Institute for Biomedical Technology and Technical Medicine, Department of Biomaterials Science and Technology, Faculty of Science and Technology, University of Twente, Enschede, The Netherlands; Department of Biomedical Engineering, University Medical Centre Groningen, Groningen, The Netherlands*

ELIA A. GUZZI • *Macromolecular Engineering Laboratory, Department of Mechanical and Process Engineering, ETH Zürich, Zürich, Switzerland*

JULES HARINGS • *Aachen-Maastricht Institute for Biobased Materials (AMIBM), Maastricht University, Brightlands Chemelot Campus, Geleen, The Netherlands*

YAN YAN SHERY HUANG • *Department of Engineering, University of Cambridge, Cambridge, UK*

YANG-HEE KIM • *Bone and Joint Research Group, Centre for Human Development, Stem Cells and Regeneration, Institute of Developmental Sciences, University of Southampton, Southampton, UK*

XIA LI • *Department of Engineering, University of Cambridge, Cambridge, UK*

ZHAOYING LI • *Department of Engineering, University of Cambridge, Cambridge, UK*

KAZIM K. MONCAL • *Engineering Science and Mechanics Department, Penn State University, University Park, PA, USA; The Huck Institutes of the Life Sciences, Penn State University, University Park, PA, USA*

LORENZO MORONI • *Department of Complex Tissue Regeneration, MERLN Institute for Technology Inspired Regenerative Medicine, Maastricht, The Netherlands*

CARLOS MOTA • *Department of Complex Tissue Regeneration, MERLN Institute for Technology Inspired Regenerative Medicine, Maastricht, The Netherlands*

PAMELA MOZETIC • *Università Campus Bio-Medico di Roma, Rome, Italy; Center for Translational Medicine, International Clinical Research Center, St. Anne's University Hospital (FNUSA-ICRC), Brno, Czechia*

DEANNA NICDAO • *Institute of Health and Biomedical Innovation, Queensland University of Technology, Brisbane, QLD, Australia*

RICHARD O. C. OREFFO • *Bone and Joint Research Group, Centre for Human Development, Stem Cells and Regeneration, Institute of Developmental Sciences, University of Southampton, Southampton, UK*

IBRAHIM T. OZBOLAT • *Engineering Science and Mechanics Department, Penn State University, University Park, PA, USA; The Huck Institutes of the Life Sciences, Penn State University, University Park, PA, USA; Biomedical Engineering Department, Penn State University, University Park, PA, USA; Materials Research Institute, Penn State University, University Park, PA, USA*

DARIO PUPPI • *BIOLab Research Group, Department of Chemistry and Industrial Chemistry, University of Pisa, UdR INSTM Pisa, Pisa, Italy*

HÉLOÏSE RAGELLE • *The David H. Koch Institute for Integrative Cancer Research, Massachusetts Institute of Technology, Cambridge, MA, USA*

ALBERTO RAINER • *Università Campus Bio-Medico di Roma, Rome, Italy*

ALFREDO RONCA • *Institute of Polymers, Composites and Biomaterials - National Research Council (IPCB-CNR), Naples, Italy*

SARA RONCA • *Department of Materials, Holywell Park, Loughborough University, Leicestershire, UK*

DROR SELIKTAR • *Faculty of Biomedical Engineering, TECHNION Israel Institute of Technology, Haifa, Israel*

WENMIAO SHU • *Institute of Mechanical Process and Energy Engineering, Heriot-Watt University, Edinburgh, UK; Department of Biomedical Engineering, University of Strathclyde, Glasgow, UK*

RAVI SINHA • *Department of Complex Tissue Regeneration, MERLN Institute for Technology Inspired Regenerative Medicine, Maastricht, The Netherlands*

BINGJIE SUN • *Department of NanoEngineering, University of California, San Diego (UCSD), La Jolla, CA, USA*

WOJCIECH SWIESZKOWSKI • *Warsaw University of Technology, Warsaw, Poland*

ATABAK GHANIZADEH TABRIZ • *Institute of Mechanical Process and Energy Engineering, Heriot-Watt University, Edinburgh, UK; Department of Biomedical Engineering, University of Strathclyde, Glasgow, UK*

MARK W. TIBBITT • *Macromolecular Engineering Laboratory, Department of Mechanical and Process Engineering, ETH Zürich, Zürich, Switzerland*

YANG WU • *Engineering Science and Mechanics Department, Penn State University, University Park, PA, USA; The Huck Institutes of the Life Sciences, Penn State University, University Park, PA, USA; School of Mechanical Engineering and Automation, Harbin Institute of Technology, Shenzhen, China*

CLAIRE YU • *Department of NanoEngineering, University of California, San Diego (UCSD), La Jolla, CA, USA*

YIN YU • *Institute for Synthetic Biology, Shenzhen Institutes of Advanced Technology, Chinese Academy of Sciences, Shenzhen, China*

WEI ZHU • *Department of NanoEngineering, University of California, San Diego (UCSD), La Jolla, CA, USA*

Part I

Design Principles

Chapter 1

Biomimetic Boundary-Based Scaffold Design for Tissue Engineering Applications

Henrique A. Almeida and Paulo J. Bártolo

Abstract

The design of optimized scaffolds for tissue engineering and regenerative medicine is a key topic of current research, as the complex macro- and micro-architectures required for scaffold applications depend not only on the mechanical properties but also on the physical and molecular queues of the surrounding tissue within the defect site. Thus, the prediction of optimal features for tissue engineering scaffolds is very important, for both its physical and biological properties.

The relationship between high scaffold porosity and high mechanical properties is contradictory, as it becomes even more complex due to the scaffold degradation process. Biomimetic design has been considered as a viable method to design optimum scaffolds for tissue engineering applications. In this research work, the scaffold designs are based on biomimetic boundary-based bone micro-CT data. Based on the biomimetic boundaries and with the aid of topological optimization schemes, the boundary data and given porosity is used to obtain the initial scaffold designs. In summary, the proposed scaffold design scheme uses the principles of both the boundaries and porosity of the micro-CT data with the aid of numerical optimization and simulation tools.

Key words Computational technologies, Topological optimization, Tissue engineering, Scaffolds, Micro-CT data

1 Introduction

In tissue engineering, the formation of tissues with desirable properties strongly relies on the mechanical properties of the scaffolds at a macroscopic and microscopic level. Macroscopically, the scaffold must bear loads to provide stability to tissues while it is being formed fulfilling its volume maintenance function. At the microscopic level, both cell growth and differentiation and ultimate tissue formation are dependent on the mechanical input to cells. Thus, the scaffold must be able to withstand specific loads and transmit them in an appropriate way to the growing and surrounding cells and tissues [1–3].

Alberto Rainer and Lorenzo Moroni (eds.), *Computer-Aided Tissue Engineering: Methods and Protocols*,
Methods in Molecular Biology, vol. 2147, https://doi.org/10.1007/978-1-0716-0611-7_1,

The design of optimized scaffolds for tissue engineering is a key topic of research, as the complex macro- and micro-architectures required for a scaffold depend on the mechanical properties and physical and molecular queues of the surrounding tissue at the defect site. One way to achieve such hierarchical designs is to create a library of unit cells (the scaffold is assumed to be a "Lego" structure), which can be assembled through a specific computational tool [4–6].

In this work, a topological optimization strategy is presented to find out the best material distribution used for a construct subject to either a single load or a multiple load distribution, maximizing its mechanical behavior under tensile and shear stress solicitations. The proposed topological optimization scheme enables the design of ideal topological architectures based on existing biologic micro-CT data for the design of biomimetic scaffolds.

2 Engineering Optimization Schemes

The classical problem in engineering design consists in finding the optimum geometric configuration of a structure that maximizes a given cost objective function with boundary conditions and constraints. Structural optimization can be classified as follows [7, 8]: size optimization, shape optimization, and topological optimization.

In size optimization, only the cross section of a structure is optimized. A typical size feature of a given structure, such as the thickness of a beam, is either increased or decreased in order to improve its performance. In shape optimization, the shape of the structure is obtained by changing the shape of the used components with other components of different shapes, in order to improve a desired variable within a system. In topological optimization, the shape and connectivity of the domain are both design variables.

Topological optimization provides the first design concept of the structure's material distribution. Its goal is to minimize the structure compliance while satisfying the constraints of volume removal. As the structure compliance is twice the strain energy, the objective function of minimizing structure compliance is equivalent to minimizing strain energy [9, 10].

In spite of several attempts to define optimized scaffolds [11–16], there is no work correlating both porosity and mechanical properties with topological information. Scaffolds must be highly porous structures but also effective from a mechanical point of view. This is a complex issue, fundamental for tissue engineering applications and not yet fully addressed. This paper proposes an optimized strategy to obtain scaffolds with an appropriate topology maximizing both porosity and mechanical behavior. The methodology proposed in this paper is of simple implementation and does not require high computational calculation time.

Topological optimization, aiming to find the best use of material according to a "maximum-stiffness" design, requires neither parameters nor the explicit definition of optimization variables. The objective function is predefined, as are the state variables (constrained dependent variables) and the design variables (independent variables to be optimized). The topological optimization problem requires the problem definition (material properties, model and loads), the objective function (the function to be minimized or maximized), and the state variables corresponding to the percentage of material to be removed [8–10, 17–21].

From a mechanical point of view, the goal of topological optimization is to minimize the total compliance, which is proportional to the strain energy. Figure 1 illustrates the general topological optimization scheme considered in this work.

The design variables are internal, pseudo-densities that are assigned to each finite element in the topological problem. The pseudo-density for each element varies from 0 to 1, where $\eta_i \approx 0$ represents material that is to be removed and $\eta_i \approx 1$ represents material that must be maintained.

For a given domain $\Omega \subseteq \mathcal{R}^2(\mathcal{R}^3)$, regions $\Omega(\Gamma_t)$ and fixed boundaries, the optimization goal is to find the optimal elasticity tensor $E_{ijkl}(x)$, which takes the form [20, 21]:

$$E_{ijkl}(x) = \eta(x)\overline{E}_{ijkl} \tag{1}$$

where $\overline{E}_{ijkl}$ is the constant rigidity tensor for the considered material and $\eta(x)$ is an indicator function for a region $\Omega^* \subseteq \Omega$ that is occupied by material:

$$\eta(x) = \begin{cases} 1, & if \quad x \in \Omega^* \\ 0, & if \quad x \notin \Omega^* \end{cases} \tag{2}$$

Considering the energy bilinear form:

$$a(u, v) = \int_{\Omega} \sum_{i,j,k,l=1}^{3} E_{ijkl}\varepsilon_{ij}(u)\varepsilon_{kl}(v)dx \tag{3}$$

with linearized strains:

$$\varepsilon_{ij} = \frac{1}{2}\left(\frac{\partial u_i}{\partial x_j} + \frac{\partial u_j}{\partial x_i}\right), \quad i,j = 1,2,3 \tag{4}$$

and the load linear form:

$$l(u) = \int_{\Omega} fu \ dx + \int_{\Gamma_t} tu \ ds \tag{5}$$

The optimization problem considered here is defined as follows:

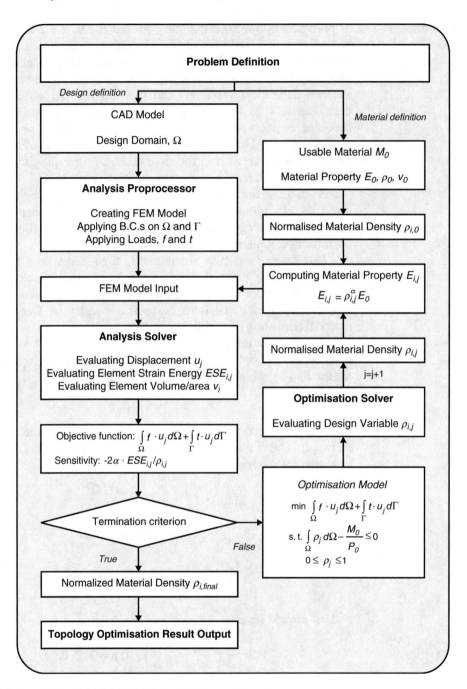

Fig. 1 A general topological optimization process

$$\min_{\eta^*(x_\varepsilon),\, i=1,\dots,n} l(u) \;=\; \int_\Omega fu \; dx + \int_{\Gamma_t} tu \; ds \tag{6}$$

subjected to constraints:

$$(i) \qquad \int_{\Omega} \sum_{i,j,k,l=1}^{3} \widetilde{E}_{ijkl}(\eta(x))\varepsilon_{ij}(u(x))\varepsilon_{kl}(v(x))dx = l(v) \qquad \forall v \in U$$

$$(ii) \qquad 0 \leq \eta(x) \leq 1 \qquad \forall x \in \Omega$$

$$(iii) \qquad \int_{\Omega} \eta(x)dx \leq Vol$$

2.1.1 Finite Element Discretization for the Optimization Problem

The domain Ω is represented as a collection of a finite number of subdomains. This is called domain discretization. Each subdomain is called a finite element and the collection of elements is called the finite element mesh. In this case, $\eta(x)$ was discretized by assigning a constant value on each element of the finite element model, establishing a constant function $\eta^*(x)$ to approximate $\eta(x)$.

2.1.2 Topological Optimization Algorithm

The algorithm considered to obtain the solution of the minimum compliance problem is based on the following update strategy [22]:
For $e = 1, \ldots, n$:

$$\eta_e^{k+1} = \begin{cases} \max\left\{(1-\zeta)\eta_e^k, eps\right\} & , \text{if } \eta_e^k\left(D_e^{k+1}\right)^\alpha \leq \max\left\{(1-\zeta)\eta_e^k, eps\right\} \\ \eta_e^k\left(D_e^{k+1}\right)^\alpha & , \text{if } \max\left\{(1-\zeta)\eta_e^k, eps\right\} \leq \eta_e^k\left(D_e^{k+1}\right)^\alpha \leq \min\left\{(1+\zeta)\eta_e^k, 1\right\} \\ \min\left\{(1+\zeta)\eta_e^k, 1\right\} & , \text{if } \min\left\{(1+\zeta)\eta_e^k, 1\right\} \leq \eta_e^k\left(D_e^{k+1}\right)^\alpha \end{cases}$$

$$(7)$$

with an appropriate weighing factor α, a move limit ζ, and an upper limit eps > 0. To perform the update strategy in Eq. 7 for a given data η_e^k, $e = 1, \ldots, n$, eps, ζ, α, it is necessary first to compute D_e^{k+1}, $e = 1, \ldots, n$ which is given by the following equation:

$$D_e^{k+1} = \left(\Lambda^{k+1}\right)^{-1} E \sum_{i,j,k,l=1}^{3} \frac{\partial \widetilde{E}_{ijkl}\left(\eta_e^k\right)}{\partial \eta} \varepsilon_{ij}^k(u(x_e))\varepsilon_{kl}^k(u(x_e)) \qquad (8)$$

3 Topological Results and Discussion

This approach is based on μCT data of real biological tissues to create the loading and constraint surfaces of the scaffold during the topological optimization process. The goal of this approach is to obtain biomimetic optimized elements. In order to perform this kind of optimization, a trabecular bone region is considered. The corresponding STL model is shown in Fig. 2a. The STL file model obtained from the μCT data is analyzed, and nonvalid triangles are removed and errors (overlapping, degenerated triangles, gaps, etc.) corrected (Fig. 2b). Once analyzed and corrected, modelling planes are created in order to define the scaffold's element boundary space (Fig. 3). The following step involves the intersection between the

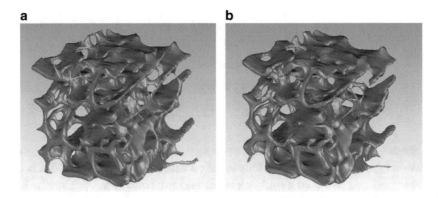

Fig. 2 (**a**) μCT STL file and (**b**) STL file after removing the triangular imperfections and errors

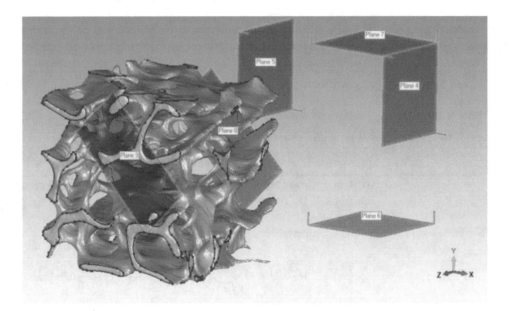

Fig. 3 Definition of the modelling planes on the STL model limiting the scaffold's boundary

modelling planes and the STL model to define the loading and constraint curves (Fig. 4) as well as the surfaces/regions for optimization (Fig. 5). Figure 6 illustrates the final scaffold unit that is considered for optimization.

The porosity of the original μCT model within the STL file is 84%, being this the considered value for simulation purposes. The block element is assumed to have a linear elastic behavior, so strain values of 0.1 are simulated by imposing a corresponding displacement according to the strain direction considered (0.2 mm along both X and Y directions and 0.184 along the Z direction). A solid tetrahedral mesh is considered for the numerical simulations (Fig. 7).

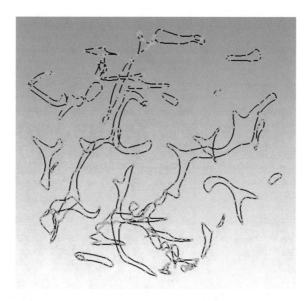

Fig. 4 Illustration of the curves obtained from the intersection between the modelling planes and the STL model

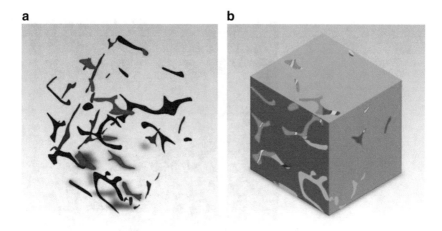

Fig. 5 Illustration of the (**a**) loading and constraint surfaces and (**b**) free non-solicited surfaces

In order to perform the topological optimization simulations, loading and constraint solicitations must be defined. In this particular study, a combined tensile and shear stress study is performed. For a better understanding of the tensile and shear stress solicitations, an individual example of each solicitation scenario is described below.

For the numerical computation of the tensile stress (Fig. 8a), a uniform displacement in a single direction is considered (the X direction), which is equivalent to the strain on the same direction (εx), imposed to a face of the block (Face A). The opposite face (Face B) of the scaffold unit is constrained and unable to have any displacement.

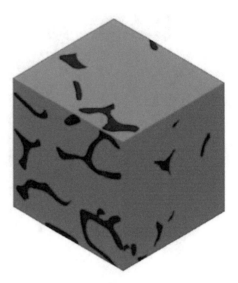

Fig. 6 Illustration of the scaffold block element considered for the topological simulations. Red, regions subjected to either loading or constraint conditions; Green, regions free of either loading or constraint conditions

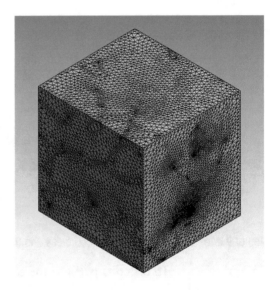

Fig. 7 Illustration of the meshed model of the scaffold

For the numerical computation of the shear stress (Fig. 8b), a uniform displacement in a single direction is considered (the Y direction), which is equivalent to the strain on the same direction (γxy), imposed to a face of the block (Face B). The opposite face (Face A) of the scaffold unit is constrained and unable to have any displacement. The two lateral faces (Faces C) are also constrained and unable to have any displacement in the X direction.

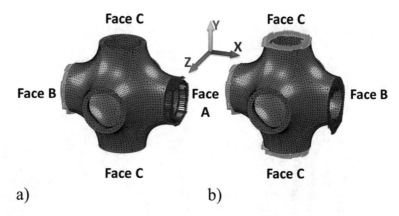

Fig. 8 Loads and constraints for the numerical analysis of scaffolds under a (**a**) tensile and (**b**) shear solicitation

Considering that the scaffold units are also cubic units, the tensile solicitations only use two faces of the cube and the shear solicitation uses four faces of the cube, and it is possible to combine both tensile and shear stresses to the same cube, in other words providing the scaffold a combined tensile and shear stress solicitation.

According to the explanation of combined solicitation environment, the following three scenarios are considered (Fig. 9):

- Scenario 1: specific regions of the block faces are submitted to tensile stress in the XX direction (ε_{xx}) and a shear stress in the clockwise direction (γ_{yz}), and the remaining specific regions are constrained (XX axis rotation).

- Scenario 2: specific regions of the block faces are submitted to tensile stress in the YY direction (ε_{yy}) and a shear stress in the clockwise direction (γ_{xz}), and the remaining specific regions are constrained (YY axis rotation).

- Scenario 3: specific regions of the block faces are submitted to tensile stress in the ZZ direction (ε_{zz}) and a shear stress in the clockwise direction (γ_{xy}), and the remaining specific regions are constrained (ZZ axis rotation).

In Fig. 9, the yellow regions are subjected to loading conditions, while the blue regions are constraint regions. The gray regions are free of either loading or constraint conditions.

The obtained results are:

- Scenario 1: Figure 10a presents a valid topological scaffold model in two different positions.

- Scenario 2: Figure 10b presents a valid topological scaffold model in two different positions.

- Scenario 3: Figure 10c presents a valid topological scaffold model in two different positions.

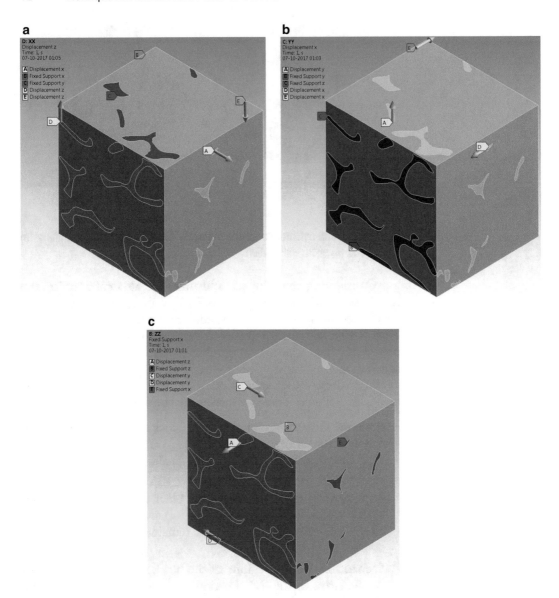

Fig. 9 Illustration of the variation of the solicitations according to each direction: (**a**) tensile and rotational displacement in the *XX* direction; (**b**) tensile and rotational displacement in the *YY* direction; (**c**) tensile and rotational displacement in the *ZZ* direction

Figure 10 illustrates the topological results for each case study. These results allow us to conclude that this approach enables to produce more biomimetic topologies. The initial external topology of the μCT data, which corresponds to a structure of a well-defined mechanical behavior and porosity level, is the starting point of an optimization scheme that enables to obtain constructs with different levels of porosity and mechanical properties according to the required applications. From the results it is also possible to observe that the inner core of the scaffold is left hollow. In order to force the

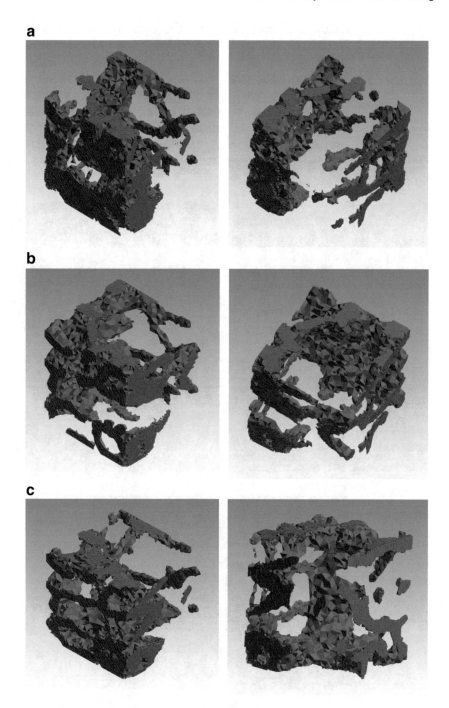

Fig. 10 Illustration of the topological optimization results according to each displacement direction: (**a**) DX, displacement in the *XX* direction; (**b**) DY, displacement in the *YY* direction; (**c**) DZ, displacement in the *ZZ* direction

creation of material toward the internal core of the scaffolds, an internal oval-shaped unit was defined and introduced to the initial CAD models. The initial case scenarios are maintained with an additional fixed oval body internally. This constraint forcing the creation of material, not only in the boundary of the scaffolds but also internally to the scaffold, increases its structural integrity.

Similarly, to the previous three scenarios, the following three present the same loading and constraint solicitations along with an internal fixed core structure (Fig. 11):

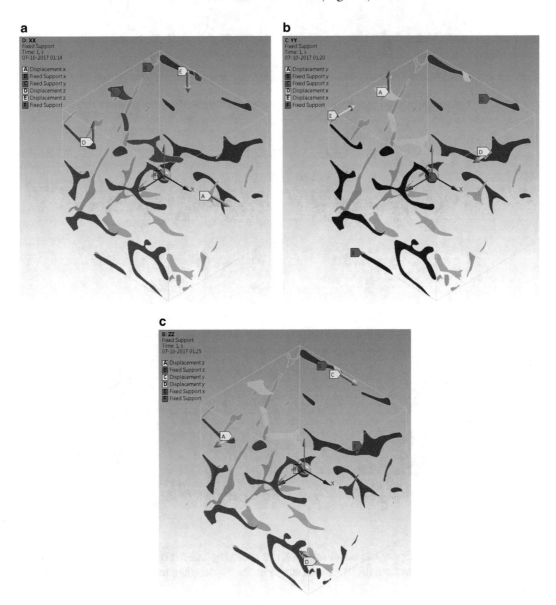

Fig. 11 Illustration of the variation of the solicitations according to each direction: (**a**) tensile and rotational displacement in the *XX* direction; (**b**) tensile and rotational displacement in the *YY* direction; (**c**) tensile and rotational displacement in the *ZZ* direction

- Scenario 1: specific regions of the block faces are submitted to tensile stress in the XX direction (ε_{xx}) and a shear stress in the clockwise direction (γ_{yz}), and the remaining specific regions are constrained (XX axis rotation).

- Scenario 2: specific regions of the block faces are submitted to tensile stress in the YY direction (ε_{yy}) and a shear stress in the clockwise direction (γ_{xz}), and the remaining specific regions are constrained (YY axis rotation).

- Scenario 3: specific regions of the block faces are submitted to tensile stress in the ZZ direction (ε_{zz}) and a shear stress in the clockwise direction (γ_{xy}), and the remaining specific regions are constrained (ZZ axis rotation).

The obtained results are:

- Scenario 1: Figure 12a presents a valid topological scaffold model in two different positions.

- Scenario 2: Figure 12b presents a valid topological scaffold model in two different positions.

- Scenario 3: Figure 12c presents a valid topological scaffold model in two different positions.

Figure 12 illustrates the topological results for each case study. Similar to the previous results, these results allow us to conclude that this type of scaffold design approach enables to produce more biomimetic topologies. In these particular scenarios, the obtained geometries now present a higher amount of material internally. In this case, the scaffolds integrity was lost in the boundaries of the structures. The goal is not to obtain similar structures as the starting model, but instead to use the starting configuration to produce novel models with more biomimetic characteristics. It is also possible to observe that the numerical solutions also need to improve their algorithms when applied to scaffold designs for tissue engineering applications.

4 Discussion

One of the existing computer-based techniques for scaffold design is topological optimization. The goal of topological optimization is to find the best use of material for a body that is subjected to either a single load or a multiple load distribution, maximizing its mechanical behavior under tensile and shear stress solicitations. The initial external topology of the μCT data, which corresponds to a structure of a well-defined mechanical behavior and porosity level, is the starting point of an optimization scheme that enables to obtain constructs with different levels of porosity and mechanical properties, according to the required applications, based on

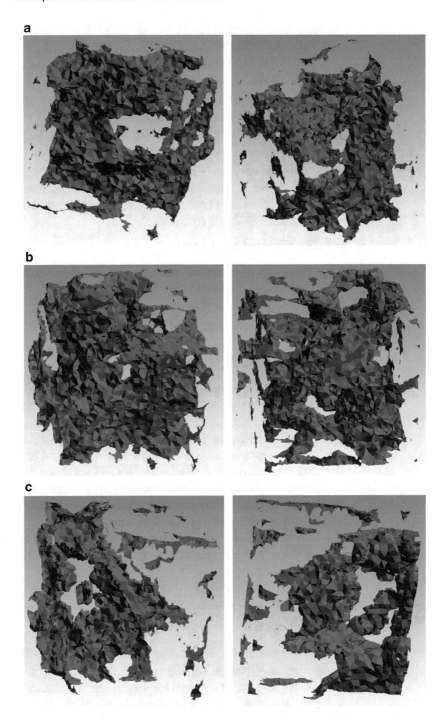

Fig. 12 Illustration of the topological optimization results according to each displacement direction: (**a**) DX, displacement in the *XX* direction; (**b**) DY, displacement in the *YY* direction; (**c**) DZ, displacement in the *ZZ* direction

biomimetic external surfaces. This particular topological optimization scheme uses the surface boundaries to produce novel models with different characteristics, which are different from the initial µCT models, enabling to design valid biomimetic scaffold topologies for tissue engineering applications.

Considering bone tissue engineering applications, from the obtained results, it is now possible to fabricate a bone tissue scaffold based on the obtained designs. Once placed within the host, these scaffolds will present a higher mechanical strength with higher porosity allowing higher and quicker bone ingrowth within the scaffold. Regarding the geometric aspect of the bone scaffold, its enhanced geometric shape allows a better adaptation of the scaffold toward the host tissue and also allows to transfer both stresses and strains along the scaffold and bone for a higher osseointegration of the bone scaffold.

Acknowledgments

The authors also wish to thank Prof. Anath Fischer from the Technion University in Haifa for supplying the micro-CT data.

References

1. Rainer A, Giannitelli SM, Accoto D, Porcellinis S, Guglielmelli E, Trombetta M (2012) Load-adaptive scaffold architecturing: a bioinspired approach to the design of porous additively manufactured scaffolds with optimized mechanical properties. Ann Biomed Eng 40(4):966–975

2. Cheah CM, Chua CK, Leong KF, Cheong CH, Naing MW (2004) Automatic algorithm for generating complex polyhedral scaffold structures for tissue engineering. Tissue Eng 10 (3–4):595–610

3. Hollister SJ (2005) Porous scaffold design for tissue engineering. Nat Mater 4(7):518–524

4. Adachi T, Osako Y, Tanaka M, Hojo M, Hollister SJ (2006) Framework for optimal design of porous scaffold microstructure by computational simulation of bone regeneration. Biomaterials 27(21):3964–3972

5. Bucklen B, Wettergreen W, Yuksel E, Liebschner M (2008) Bone-derived CAD library for assembly of scaffolds in computer-aided tissue engineering. Virtual Phys Prototyp 3(1):13–23

6. Naing MW, Chua CK, Leong KF (2008) Computer aided tissue engineering scaffold fabrication. In: Bidanda B, Bártolo PJ (eds) Virtual Prototyping & Bio-Manufacturing in Medical Applications. Springer, New York, NY

7. Lee TH (2007) Shape optimization. In: Arora JS (ed) Optimisation of structural and mechanical Systems. World Scientific, Singapore

8. Hsu MH, Hsu YL (2005) Generalization of two and three-dimensional structural topology optimization. Eng Optim 37:83–102

9. Kruijf N, Zhou S, Li Q, Mai YW (2007) Topological design of structures and composite materials with multiobjectives. Int J Solids Struct 44(22–23):7092–7109

10. Rozvany GIN (2001) Aim, scope, methods, history and unified terminology of computer-aided topology optimization in structural mechanics. Struct Multidiscipl Optim 21:90–108

11. Podshivalov L, Fischer A, Bar-Yoseph PZ (2008) Multi-scale finite-element analysis as a base for 3D computerized virtual biopsy system. In: Halevi Y, Fischer A (eds) Proceedings of the 9th biennial ASME conference on engineering systems design and analysis (ESDA2008). American Society of Mechanical Engineers, New York City

12. Podshivalov L, Fischer A, Bar-Yoseph PZ (2009) Multiresolution geometric meshing for multiscale finite element analysis of bone micro-structures as a base for 3D computerized virtual biopsy system. In: Bártolo PJ, et al.

(eds) Proceedings of the ICTE 2009 International conference on tissue engineering. IST Press, Lisbon, pp 317–323

13. Holdstein Y, Fischer A (2008) Modelling micro-scaffold-based implants for bone tissue engineering. In: Halevi Y, Fischer A (eds) Proceedings of the 9th biennial ASME conference on engineering systems design and analysis (ESDA2008). American Society of Mechanical Engineers, New York City

14. Wettergeen MA, Bucklen BS, Liebschner MAK, Sun W (2008) CAD assembly process for bone replacement scaffolds in computer-aided tissue engineering. In: Bidanda B, Bártolo PJ (eds) Virtual prototyping & bio-manufacturing in medical applications. Springer, New York, NY

15. Wettergreen MA, Bucklen BS, Starly B, Yuksel E, Sun W, Liebschner MAK (2005) Creation of a unit block library of architectures for use in assembled scaffold engineering. Comput Aided Des 37(11):1141–1149

16. Wettergreen MA, Bucklen BS, Sun W, Liebschner MAK (2005) Computer-aided tissue engineering of a human vertebral body. Ann Biomed Eng 33(10):1333–1343

17. Neches L, Cisilino A (2008) Topology optimization of 2D elastic structures using boundary elements. Eng Anal Bound Elem 32:533–544

18. Ansola R, Veguería E, Canales J, Tárrago J (2007) A simple evolutionary topology optimization procedure for compliant mechanism design. Finite Elem Anal Des 44:53–62

19. Bendsøe MP, Sigmund O (2003) Topology optimization: theory, methods and applications. Springer, New York, NY

20. Mlejnek HP, Schirrmacher R (1993) An engineering approach to optimal material distribution and shape finding. Comput Method Appl Mech Eng 106:1–26

21. Mlejnek HP (1992) Some aspects of the genesis of structures. Struct Multidiscipl Optim 5:64–69

22. Vogel F (1997) Topology Optimisation of linear-elastic Structures with ANSYS 5.4. NAFEMS Conference on Topology Optimisation, Aalen, Germany

Chapter 2

Triply Periodic Minimal Surfaces (TPMS) for the Generation of Porous Architectures Using Stereolithography

Sebastien B. G. Blanquer and Dirk W. Grijpma

Abstract

A new generation of sophisticated tissue engineering scaffolds are developed using the periodicity of trigonometric equations to generate triply periodic minimal surfaces (TPMS). TPMS architectures display minimal surface energy that induce typical pore features and surface curvatures. Here we described a series of TPMS geometries and developed a procedure to build such scaffolds by stereolithography using biocompatible and biodegradable photosensitive resins.

Key words Triply periodic minimal surfaces (TPMS), Minimal surface energy, Trigonometric equations, Stereolithography, Scaffold, Biodegradable polymer, Photosensitive resins

1 Introduction

Scaffold design and pore geometries play a major role in cell and tissue organization in tissue-engineered constructs. It is becoming clear that the high biological and functional complexity of the human tissues with specific (micro)architecture and vascular networks may require more specific and sophisticated scaffold geometries using appropriate multifunctional materials. The expansion of additive manufacturing technologies in the development of sophisticated scaffolds remains in constant evolution. Among the existing additive manufacturing techniques, stereolithography (SL) is recognized for its remarkable efficiency and considerable advantages in terms of versatility in manufacturing, high accuracy, and quickness. It offers unique ways to precisely control substrate architecture [1, 2]. The principle of the fabrication is based on a spatially controlled solidification by photopolymerization of monomer or prepolymer resins, in liquid or viscous state, using a single-photon source. In order to build the desired 3D structures in a layer-by-layer manner with SL, it is necessary to initially design the 3D object from a "3D computer-assisted design" (CAD) file.

Alberto Rainer and Lorenzo Moroni (eds.), *Computer-Aided Tissue Engineering: Methods and Protocols*,
Methods in Molecular Biology, vol. 2147, https://doi.org/10.1007/978-1-0716-0611-7_2,
© Springer Science+Business Media, LLC, part of Springer Nature 2021

In order to increase the diversity of structure and the complexity of internal porous architecture, a new approach of CAD file conception can be used. In this optic, we developed a new generation of sophisticated, stable, and highly interconnected porous scaffolds based on triply periodic minimal surfaces (TPMS) (Fig. 1) [3]. TPMS architectures are infinite and periodic in the 3D Euclidean space and belong to the interesting class of minimal surfaces [4–6]. Minimal surfaces are frequently encountered in nature and play an essential role in guiding chemical, biochemical, and cellular processes [7–10]. The natural organization of minimal surface structures responds to physical principle that governs the forms and the motions of objects, the principle of free energy minimization. In nature and man-made environments, the systems normally try to arrange themselves to minimize their potential energy in order to consume less energy and leading to a better stability [11]. Consequently, the term "minimal surface" is directly linked to the surface energy and represents the lowest possible potential energy that a surface can have if its energy is proportional to the surface area. In addition, the minimization of the surface energy leads automatically to typical curved structure with respect of the initial fixed boundary [6, 12, 13].

The study of TPMS architecture scaffolds allows to give a new dimension and vision of the tissue engineering scaffold designs. Indeed, in addition to the great benefit of such structures in terms of interconnectivity, high specific tortuosity and maximization of the specific surface area, TPMS architectures based on such concept of minimal surface energy display specific surface curvature distribution, which has been proved to be also a significant parameter that can influence cell behavior and therefore tissue formation [14, 15].

SL technique is chosen to produce these complex scaffolds using two different materials, a rubber-like material poly(trimethylene carbonate) (PTMC) and a stiff material poly(D,L-lactide) (PDLLA). The synthesis of the photosensitive resins is conveniently prepared in two successive steps with high yields (Fig. 2), and the resin was successfully used by SL to develop a library of different TPMS architecture scaffolds.

2   Materials

The described protocol requires access to a chemistry wet lab with standard equipment.

1. Trimethylene carbonate (TMC). Monomers must be stored at −20 °C in sealed containers.

2. DL-Lactide (DLLA). Monomers must be stored at −20 °C in sealed containers.

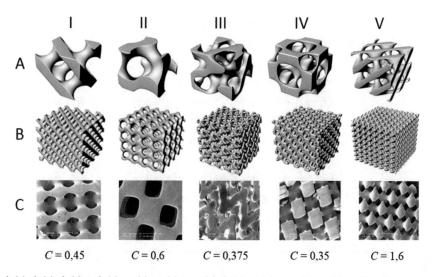

I/ $\sin(x).\sin(y).\sin(z) + \sin(x).\cos(y).\cos(z) + \cos(x).\sin(y).\cos(z) + \cos(x).\cos(y).\sin(z) = C$

II/ $\cos(x).\sin(y) + \cos(y).\sin(z) + \cos(z).\sin(x) = C$

III/ $\cos(2x).\sin(y).\cos(z) + \cos(2y).\sin(z).\cos(x) + \cos(2z).\sin(x).\cos(y) = C$

IV/ $8.\cos(x).\cos(y).\cos(z) + 1.(\cos(2x).\cos(2y).\cos(2z)) - 1.(\cos(2x).\cos(2y) + \cos(2y).\cos(2z) + \cos(2z).\cos(2.x)) = C$

V/ $1.(\sin(2x).\sin(2y) + \sin(2y).\sin(2z) + \sin(2x).\sin(2z)) + 1.(\cos(2x).\cos(2y).\cos(2z)) = C$

Fig. 1 Visualization of five TPMS scaffolds with their respective harmonic trigonometric functions. (**a**) CAD designs representing the repeated unit cells, (**b**) assemblies of 4 × 4 × 4 unit cells to create a sophisticated 3D porous structure, (**c**) SEM images of the built structures prepared from the PTMC resin by SL. Offset values C are chosen for each structure to reach 65% of porosity

Fig. 2 Synthesis of PTMC photosensitive resin by successive ROP and end-functionalization

3. Tin 2-ethylhexanoate 92.5–100.0% (SnOct2).

4. Trimethylolpropane ≥98.0% (TMP).

5. Trimethylamine ≥99% (NEt3).

6. Methacrylic anhydride 94% (contains 2000 ppm Topanol A as inhibitor).

7. Methanol absolute.

8. Anhydrous dichloromethane $\geq 99\%$.

9. Propylene carbonate.

10. Lucirin TPO-L (ethyl 2,4,6-trimethylbenzoylphenyl phosphinate, BASF) photoinitiator.

11. Orasol Orange G (Ciba Specialty Chemicals) to control the degree of penetration of blue light in the resin.

12. Three-necked flask. Dry for several hours in an oven at 100 °C prior to use.

13. Crosslinking cabinet (Ultralum).

14. Stereolithographic 3D printer (Perfactory Mini Multilens, EnvisionTec).

3 Methods

3.1 Synthesis of the Biodegradable and Biocompatible Three-Armed Oligomers by Ring Opening Polymerization

Two types of biocompatible and biodegradable polymeric materials are investigated to prepare the TPMS scaffolds, materials based on poly(trimethylene carbonate) (PTMC) and poly(D,L-lactide) (PDLLA). A multi-gram production of polymer resin is afforded with predetermined molecular weights. Low chain length is preferred to facilitate the end-functionalization and the crosslinking steps. The configuration of the polymer (single arm, star-shaped, etc.) is determined by the initiator molecule, and the molecular weight can be adjusted by varying the monomer to the initiator molar ratio. Three-armed and hydroxy-terminated oligomers targeting 5000 g/mol as global chain length are synthesized. The targeted molecular weight of 5000 g/mol corresponds to an approximate degree of polymerization of 16 trimethylene carbonate units per arm and an approximate degree of polymerization of 23 lactic acid units per arm.

All manipulations must be performed in a fume hood. The experimenter must wear a lab coat and gloves during all the different chemical steps.

1. To synthesize three-armed PTMC, add TMC monomer (0.98 mol; 100 g), Sn(Oct)$_2$ catalyst (0.05 wt% of monomer), and TMP initiator (0.0196 mol; 2.62 g) (*see* **Note 1**). Add the compounds under argon into the dried flask, and apply to the mixture 10 cycles of vacuum and argon backfill at room temperature and under stirring (*see* **Note 2**).

2. Perform bulk polymerization for 48 h in argon atmosphere at 130 °C by heating with an oil bath.

3. Same protocol can be adjusted for the synthesis of poly(D,L-lactide) oligomers. Three-armed PDLLA oligomer synthesis is performed by ring opening polymerization of D,L-lactide (0.694 mol; 100 g), catalyzed by Sn(Oct)$_2$ (0.05 wt% of

monomer), and initiated by TMP (6.94 mmol; 0.93 g) at 130 °C for 48 h under an argon atmosphere.

4. The monomer conversion and the degree of polymerization can be determined by ^{1}H-NMR spectroscopy using CDCl$_3$ as a solvent. Assuming that each hydroxyl group of the glycerol initiates polymerization of TMC, the degree of polymerization is determined by comparing the peak integral that characterizes the -CH$_2$- groups of the TMC at 4.2 ppm (4H) or 2.05 ppm (2H) with the CH$_3$- groups from the TMP initiator at 0.8 ppm (3H). Monomer conversion is calculated from the peak integral of the PTMC at 4.2 ppm (4H) or 2.05 ppm (2H) compared to the peak integral of the TMC monomer at 4.4 ppm (4H) or 2.2 ppm (2H).

5. Same characterization and efficacy are obtained with PDLLA oligomers. The conversion is determined by the characteristic peak integral of lactide monomers at 5.05 ppm corresponding to the -CH- (1H) groups and the peak integral of the -CH- from PDLLA at 5.15 ppm. Conversion rate for both polymer types is traditionally close to 95–99%.

3.2 Preparation of the Photosensitive Resin Precursor

Following the synthesis of the hydroxy-terminated oligomers, the next step of the resin preparation consists of the end-functionalization of the oligomers by photosensitive groups able to polymerize under light initiation. This functionalization is a key step to achieve a photocrosslinkable resin. The most straightforward approach of such functionalization is to end-cap the polymer chains through the reaction of the hydroxyl end-groups with photoreactive acrylate or methacrylate groups in order to induce radical photopolymerization [16], which leads to photosensitive macromers. Radical photopolymerization is a fast, selective, and highly effective reaction, which is a crucial requirement in SL. In conventional SL technology, the resin composition includes the photosensitive polymer resin, a photoinitiator, and if necessary an inert or reactive diluent.

1. After dissolution of the hydroxy-terminated oligomers in anhydrous dichloromethane (100 mL) (*see* **Note 3**), add an excess of anhydrous methacrylic anhydride (0.15 mol; 22 mL) in the presence of triethylamine (0.15 mol; 20 mL). The reaction is performed at room temperature under stirring and argon atmosphere for 3 days.

2. Purify photosensitive resin by precipitation (*see* **Note 4**). Dispense the resin dropwise in 500 mL of ice-cold methanol inside a beaker under vigorous stirring. Recover and vacuum-dry the precipitate. A sticky white paste should be obtained.

3. The degree of functionalization can be determined by ^{1}H-NMR analysis in CDCl$_3$. Taking into account the degree

of polymerization determined previously, the end-functionalization rate is assessed by comparing the peak integral of the PTMC at 4.2 ppm (4H) or 2.05 ppm (2H) with the grafted methacrylate with the peak integral at 6.06 ppm (1H) and 5.51 ppm (1H) corresponding to the acrylate bonds CH_2-. Purity of the macromers is also evaluated by ^{1}H-NMR with the presence of the remained non-grafted methacrylate molecules with chemical shifts at 6.09 ppm (1H) and 5.51 ppm (1H). Degree of functionalization on PDLLA macromers is assessed via the integral peak at 5.15 ppm of the -CH- (1H) groups (*see* **Note 5**). End-functionalization of these types of macromers is 90–95% (*see* **Notes 6** and **7**).

3.3 Photocross-linking Assays Using the Synthesized Macromer Resins

Prior to any building by SL, the crosslinking ability should be assessed initially on a film. Crosslinking characterization consists in the measure of the gel content.

1. Dissolve macromers in dichloromethane (30 wt%) in order to decrease the viscosity (*see* **Note 8**). Add Lucirin TPO-L (5 wt% relative to the macromers) (*see* **Note 9**).

2. Use a mold to create films with a thickness of 500 μm, and irradiate the resin for 10 min at a wavelength of 452 nm into a crosslinking cabinet under argon flow to avoid radical polymerization quenching.

3. Gel content determination is performed by a weighing method (at least in triplicate). Vacuum-dry film and weigh it to give m_0. Rinse film in dichloromethane, and refresh solvent twice. Vacuum-dry film until a constant weight is reached (m_1). Efficient crosslinking procedure should lead to gel content above 90%. The gel content is defined as:

$$\text{Gel content (\%)} = \frac{m_0}{m_1} \times 100 \qquad (1)$$

3.4 Computer-Assisted Design Based on Triply Periodic Minimal Surface

As a rapid prototyping process, SL can create physical 3D objects from designs obtained by computer-aided design (CAD). Such CAD files can be basically generated by graphical computer software, but it can also be obtained from data acquired with medical imaging techniques such as magnetic resonance imaging (MRI) or computed tomography (CT). In this chapter, we describe another advanced approach to generate sophisticated 3D porous structures using the periodicity of trigonometric equations to generate triply periodic minimal surfaces (TPMS) (Fig. 1) [3, 17]. TPMS are mathematically defined and are recognized to be infinite and periodic in the 3D Euclidean space and display specific surface curvatures making them interesting porous architectures for highly controllable and homogeneous scaffold designs [3, 18]. As shown in Figs. 1 and 3a, TPMS structures are periodic in three

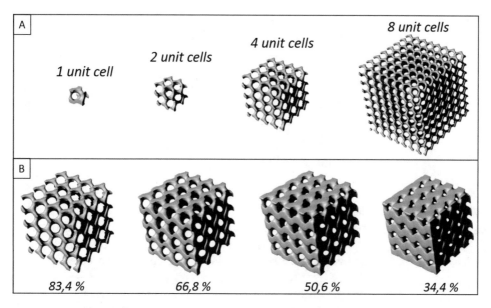

Fig. 3 Variation of the scaffolds features in terms of number of unit cells defined by the boundary conditions (**a**) and porosity controlled by the constant C (**b**)

independent directions and can be infinitely reproduced by repetition of cubic translational unit cells and thus can be used to construct a 3D porous scaffold.

1. Each TPMS design is defined by a trigonometric equation that is closely approximated by a periodic nodal surface [12] which leads to bicontinuous (or biphasic) TPMS porous structures. Figure 1 lists the approximated periodic nodal equations for five TPMS structures. Such approximation allows designing TPMS architectures, which therefore can form a solid/void interface where the void space represents the pores of the 3D scaffold and the solid space is materialized by the polymer.

2. TPMS designed scaffolds are obtained using K3dSurf v0.6.2 software (freeware from http://k3dsurf.sourceforge.net) that generates the CAD file. Boundary conditions must be adopted in the software to control the number of unit cells expected for the design scaffold (Fig. 3a). Consequently for scaffolds with four unit cells in x, y, and z dimensions, the boundary conditions are $x, y, z = [-4\pi, 4\pi]$, and a common number of 64 unit cells within the scaffold are generated. The obtained bicontinuous geometry from the TPMS structures gives a scaffold with total pore interconnectivity.

3. By modulating the linear term (C) to the nodal equation, it is possible to vary the pore characteristics (porosity and pore size) and the surface curvature (Fig. 3b). Therefore, the role of the constant C can be defined as offsetting value increasing the volume fraction of the void space. In the equations presented in

Fig. 1, the offset value has been chosen to reach 65% of porosity for all the porous structures designed and built by SL.

4. The generated CAD files can be then converted into STL files using computer-aided design software (e.g., Rhinoceros3D, McNeel). Volume scaffolds in cubic shape have been fixed to $10 \times 10 \times 10$ mm^3 for the TPMS scaffolds shown in Fig. 1.

5. EnvisionTec Perfactory RP2.0 software is used to slice the 3D CAD files in multiple layers that are sequentially fabricated by photocrosslinking.

3.5 Preparation of TPMS Scaffolds by Stereolithography

Of crucial importance in the development of precise 3D scaffold structure are not only the designed models imagined but also the types of biomaterials employed as they influence the fabrication and the properties of the fabricated 3D objects. To be used in SL, the resins should rapidly solidify upon illumination with light. In addition, SL is recognized for its high accuracy and resolution in terms of fabrication, and therefore it is necessary to develop a calibration procedure of the apparatus for each new resin batch.

1. Digital light processing (DLP) SL with a top-down approach is used to fabricate the different TPMS scaffolds. SL apparatus is equipped with a digital mirror device allowing projections of 1280×1024 pixels, each measuring 16×16 μm^2. The wavelength of the blue light irradiation ranges from 400 to 550 nm, with a peak at 440 nm.

2. For appropriate use with SL, the resin should be liquid or semi-viscous in order to ensure the transfer from liquid to a cross-linked solid. Hence, adjust resin viscosity by diluting it with a nontoxic solvent such as propylene carbonate (for both types of resin PTMC and PDLLA) in mass proportion 25 wt%.

3. In order to initiate the photopolymerization, a photoinitiator with a decomposition wavelength in adequacy with the light irradiation of the machine (*see* **Note 10**). Add Lucirin TPO-L photoinitiator (5 wt% relative to the macromer) and Orasol Orange dye (0.15 wt% relative to the macromer) to control the penetration depth of blue light. The resin is then dark orange (*see* **Note 11**).

4. Top-down approach of fabrication implies fixation of the first layer of the building to a glass support platform. The first layer is therefore solidified and attached to the platform by photo-crosslinking. After photopolymerization of the first layer, the platform is moved away from the surface, and the liquid/viscous resin should then refill the space in order to allow the curing of the second layer. The layer-by-layer procedure continues until the complete fabrication of the designed structure (*see* **Note 12**).

5. Once the resin is prepared, the machine is calibrated through a working curve in order to precisely control the depth of cure in the resin layers. This working curve is established by measuring the photocrosslinked layer thickness for several exposure time at a certain intensity of irradiation. The correlation between the exposure time and the thickness layer expected during the building by the SL can be made using an adapted equation from the Beer-Lambert equation. This procedure has been excellently described and explained in the review from Melchels et al. [2]. On average for the PTMC and PDLLA resins, the required curing time is around 50 s with a light intensity of 18 mW/cm^2 for a sequentially layer thickness of 25 μm.

6. 3D fabrication occurs in a resin bath comprising a non-UV-absorbing plate (e.g., quartz, PDMS, etc.). For a support platform of 100 cm^2, the resin bath should have a surface of at least 120 cm^2 for a global volume of resin around 25 mL (see **Note 13**). Such volume is required to make twice building of 20 cubic scaffolds (1 cm^3).

7. After building, collect scaffolds from the platform and remove non-reacted resin by several extraction cycles (see **Note 14**). Extract non-reacted resin in 100% propylene carbonate, refreshing solvent three times a day. Exchange propylene carbonate by acetone in order to facilitate the drying step with the following acetone/propylene carbonate ratios: 20:80, 50:50; 75:25, and finally 100% acetone. Vacuum-dry scaffolds for several hours (see **Note 15**).

8. Structural and textural scaffold characterizations are performed by scanning electron microscopy (SEM) and micro-computed tomography (μCT) (see **Note 16**). Pore features in terms of interconnectivity, porosity, pore distribution, and strut thickness distribution can be assessed by these visualization analyses [3, 19].

4 Notes

1. The monomers are stored at −20 °C, and consequently before the polymerization, it is advised to equilibrate the monomers batches at room temperature before opening the sealed container.

2. Ring opening polymerization must be done in rigorous anhydrous environment. The presence of water initiates the ring opening polymerization.

3. Solubilization of PTMC or PDLLA oligomers is facilitated by stirring, but even then, it may require some time for high amounts of material.

4. All the final compounds included in the scaffolds are suited for biomedical applications. Intermediary reagents and by-products are removed during the procedure of purification for each step of the synthesis.

5. Ungrafted methacrylates from methacrylate anhydride or methacrylic acid are visible in the NMR spectra at different chemical shifts than the methacrylates grafted to the polymer, which therefore allows the reaction process to be followed, and the elimination of unreacted methacrylates by precipitation.

6. The methacrylated polymers can be defined as a photoactive resin, macromer, macromonomer, or prepolymer.

   In the absence of photoinitiator, the acrylate groups are hardly susceptible to photo-curing; however it is recommended to protect the resin from light during storage and building process to prevent premature crosslinking. It is recommended to store resin batches in a freezer, especially for PDLLA resins, which are highly sensible to degradation by hydrolysis.

7. Drying procedures must be performed under vacuum at room temperature. Rising the temperature may initiate unexpected thermal crosslinking.

8. Optimal viscosity has been found around 10 Pa·s; however higher or lower viscosity can also be used. Viscosity can be adjusted by dilution in a solvent with high boiling temperature to prevent evaporation and modification of the resin concentration. Use of an apparatus equipped with heater can also be suggested to reduce the viscosity.

9. The end-functionalization reaction might lead to a yellow solution, which is due to the oxidation of the trimethylamine added in the reaction but does not impair the efficiency of the reaction.

10. Radical photopolymerization of methacrylates is known to be efficient and fast, but in case of low crosslinking efficiency or weak mechanical properties of the building object, curing can be enhanced using a reactive solvent (e.g., NVP) or crosslinker (e.g., di- or tri-acrylate).

11. Dyes such as Orange Orasol are suggested in order to avoid the over-curing of the resin into a preceding layer.

12. Radical photopolymerization is inhibited in presence of oxygen, and the reaction should be ideally maintained continuously under flow of inert gas like argon or nitrogen. This caution can be taken conveniently during the initial curing tests process using a UV cabinet but remains hard to set up during the building by SL. That is why the bottom-up approach significantly limits the contact with the environment and the photopolymerization is only little impaired by oxygen.

13. Volume of the resin may be adjusted to fit the size of the support platform.

14. Solvent extraction is a critical step as the swelling effect may damage or destroy the scaffolds. Consequently, acetone can be substituted by ethanol. Nevertheless, purification of the scaffolds by solvent extraction must be done carefully and necessarily sequentially. Supernatant from the extraction steps should be intensely orange-colored for the first steps and completely clear for the last steps.

15. In this process, the conversion of reactive groups is usually incomplete, and post-thermal curing (90 °C for 5 h) is often applied to convert unreacted methacrylate functions into the polymer network.

16. Due to the formation of a chemical network during the building by SL, the solvent is present in the network, and therefore the size of the 3D object changes upon drying. The sizes of the designs should be adjusted to the diluent concentration in order to give scaffolds with desired sizes after building, extracting, and drying.

References

1. Bartolo PJ (2011) Stereolithography: materials, processes and applications. Springer, New-York, NY. ISBN 978-0-387-92904-0

2. Melchels FPW, Feijen J, Grijpma DW (2010) A review on stereolithography and its applications in biomedical engineering. Biomaterials 31(24):6121–6130

3. Blanquer SBG, Werner M, Hannula M, Sharifi S, Lajoinie GPR, Eglin D, Hyttinen J, Poot AA, Grijpma DW (2017) Surface curvature in triply-periodic minimal surface architectures as a distinct design parameter in preparing advanced tissue engineering scaffolds. Biofabrication 9(2):025001

4. Fischer W, Koch E (1987) On 3-periodic minimal-surfaces. Z Kristallogr 179(1–4):31–52

5. Schoen AH (2012) Reflections concerning triply-periodic minimal surfaces. Interface Focus 2(5):658–668

6. Hyde S, Andersson S, Larsson Z, Blum T, Landh S, Lidin BW, Ninham BW (1997) The language of shape: the role of curvature in condensed matter: physics, chemistry and biology. Elsevier, Amsterdam

7. Mai YY, Eisenberg A (2012) Self-assembly of block copolymers. Chem Soc Rev 41 (18):5969–5985

8. Angelova A, Angelov B, Mutafchieva R, Lesieur S, Couvreur P (2011) Self-assembled multicompartment liquid crystalline lipid carriers for protein, peptide, and nucleic acid drug delivery. Acc Chem Res 44(2):147–156

9. Tenchov B, Koynova R (2012) Cubic phases in membrane lipids. Eur Biophys J Biophys 41 (10):841–850

10. Urbas AM, Maldovan M, DeRege P, Thomas EL (2002) Bicontinuous cubic block copolymer photonic crystals. Adv Mater 14 (24):1850–1853

11. Eriksson JC, Ljunggren S (1994) The mechanical surface-tension and stability of minimal surface-structures. J Colloid Interface Sci 167 (2):227–231

12. Gandy PJF, Bardhan S, Mackay AL, Klinowski J (2001) Nodal surface approximations to the P, G, D and I-WP triply periodic minimal surfaces. Chem Phys Lett 336(3–4):187–195

13. Mackay AL (1994) Periodic minimal-surfaces from finite-element methods. Chem Phys Lett 221(3–4):317–321

14. Rumpler M, Woesz A, Dunlop JWC, van Dongen JT, Fratzl P (2008) The effect of geometry on three-dimensional tissue growth. J R Soc Interface 5(27):1173–1180

15. Werner M, Blanquer SBG, Haimi SP, Korus G, Dunlop JWC, Duda GN, Grijpma DW, Petersen A (2017) Surface curvature differentially regulates stem cell migration and

differentiation via altered attachment morphology and nuclear deformation. Adv Sci 4 (2):1600347

16. Matsuda T, Mizutani M (2002) Liquid acrylate-endcapped biodegradable poly(epsilon-caprolactone-co-trimethylene carbonate). II. Computer-aided stereolithographic microarchitectural surface photoconstructs. J Biomed Mater Res 62(3):395–403

17. Schoen AH (1970) Infinite periodic minimal surfaces without self-intersections. NASA Technical report TN D-5541, Washington, DC

18. Yoo DJ (2011) Porous scaffold design using the distance field and triply periodic minimal surface models. Biomaterials 32 (31):7741–7754

19. Narra N, Blanquer SBG, Haimi SP, Grijpma DW, Hyttinen J (2015) mu CT based assessment of mechanical deformation of designed PTMC scaffolds. Clin Hemorheol Microcirc 60(1):99–108

Chapter 3

3D Printing of Functionally Graded Films by Controlling Process Parameters

Alessandra Bonfanti, Loris Domenicale, and Atul Bhaskar

Abstract

Scaffolds are often used in bioengineering to replace damaged tissues. They promote cell ingrowth and provide mechanical support until cells regenerate. Such scaffolds are often made using the additive manufacturing process, given its ability to create complex shapes, affordability, and the potential for patient-specific solutions. The success of the implant is closely related to the match of the scaffold mechanical properties to those of the host tissue. Many biological tissues show properties that vary in space. Therefore, the aim is to manufacture materials with variable properties, commonly referred to as functionally graded materials. Here we present a novel technique used to manufacture porous films with functionally graded properties using 3D printers. Such an approach exploits the control of a process parameter, without any hardware modification. The mechanical properties of the manufactured films have been experimentally tested and analytically characterized.

Key words Additive manufacturing, Functionally graded material, Graded films, G-code, Process parameter, Extrusion rate

1 Introduction

After decades of development of medical implants made of dense materials, porous scaffolds have brought excitement to the area as they afford improved mechanical performance and they promote cell ingrowth [1, 2, 3]. Scaffolds are used in tissue engineering to mimic the extracellular matrix of the body allowing regeneration of damaged or diseased tissues [4–6]. The success of cell culture depends on the internal architecture (e.g., pore size) of the scaffolds [7–9]. In case of the regeneration of multiple tissues (e.g., bone, cartilage), different cell types are necessary. Each of them operates under different in vivo conditions, and, therefore, different scaffold geometries and porosity are required to promote cell growth [10, 11]. Such scaffolds with spatially varying architecture are commonly referred to as functionally graded scaffolds [12]. While reproducing the spatial gradation of properties

Alberto Rainer and Lorenzo Moroni (eds.), *Computer-Aided Tissue Engineering: Methods and Protocols*,
Methods in Molecular Biology, vol. 2147, https://doi.org/10.1007/978-1-0716-0611-7_3,
© Springer Science+Business Media, LLC, part of Springer Nature 2021

biomimetically in medical implants to imitate the mechanical behavior of the original tissue (e.g., bone [13]), it is also important to reduce the risk of implant failure. The mismatch of mechanical properties between the prosthesis and the surrounding tissue has been identified as a major cause of loosening of orthopedic implants because the density of the host bone reduces [14, 15]—known as stress shielding. When the bone loss is significant, fracture can easily occur.

Conventional manufacturing techniques have been applied to the fabrication of lattice structures in the past. For example, fiber bonding, membrane lamination, molding, solvent casting, phase separation, and foaming have been used in the context [16, 17]. However, these techniques do not allow control of the scaffold architecture—i.e., possible tailoring of structural parameters such as pore size and pore network; and they are time and cost demanding [18]. The advent of additive manufacturing has enabled the precise manufacture of lattice structures, thanks to the possibility of controlling the spatial deposition of material. Such a method allows an accurate control of pore size and structure, while being cost-efficient. The popularity of additive manufacturing has further increased given the possibility of manufacturing complex shapes.

Scaffolds with functionally graded porosity are commonly manufactured using additive manufacturing techniques. One possible way to obtain scaffolds with variable stiffness is to engineer the spacing of the lattice structure: smaller pore size is used where higher stiffness is required [19]. However, this strategy does not work on woodpile lattices, as the bending stiffness of such lattices is independent of the spacing across the bending direction. Here we propose a solution that involves the deposition of filaments with continuously variable cross section. Currently available technologies allow printing filaments with variable stiffness by using variable cross-section nozzles [20]. However, this solution requires specific hardware modifications. Here we present a manufacturing solution to fabricate structured materials with variable stiffness via the control of a process parameter. This technique allows a continuous variation of the stiffness by controlling the diameter of the extruded filaments point to point without requiring a hardware modification.

2 Materials

Fused deposition modelling is an affordable method used to manufacture biomedical scaffolds. Several 3D printers are currently available on the market. Here we make use of the Ultimaker 2 commercial machine [21].

<table>
<tr><td>

2.1 Printing Equipment

</td><td>

1. Ultimaker 2 printer (Ultimaker).
2. Polylactic acid filament (*see* **Note 1**).

</td></tr>
<tr><td>

2.2 Characterization Equipment

</td><td>

1. Universal tensile tester equipped with compression plates.
2. Optical microscope equipped with a digital camera.

</td></tr>
</table>

3 Methods

3.1 G-Code Instructions

The printer-head movements are controlled by the G-code—a programming language originally developed to control the tool path of CNC machines. The commands starting with "G" instruct the machine as to what the tool path must obey:

G10: the machine retracts the filaments from the printing head and it stops extruding. This command can be followed by the option *S0* for "short retract" or *S1* for "long retract."

G11: the machine pushes the filament into the printing head and it starts extruding.

G1: the machine linearly moves to the specified point while extruding the melted material. This command is followed by five parameters *F*, *X*, *Y*, *Z*, and *E*. The printing head moves to the point of coordinates *X*, *Y*, and *Z*, while the spool filament is pushed into the printing head at a speed *F* (mm/min). The parameter *F* is often referred to as feed rate. If *X*, *Y*, and *Z* are not specified, the machine will automatically assign the latest values; therefore, the machine will not move. If *F* is not specified, the machine uses the last assigned value. *E* is the volume of material being extruded. For Ultimaker 2, this parameter is a cumulative value; thus for each movement, the value is increased by the volume of the material extruded during the current movement. If the printing head of diameter d_{nozzle} moves from point $A(x_1, y_1)$ to $B(x_2, y_2)$, the value of *E* is increased by

$$\Delta E = A_{\text{nozzle}} \overline{AB} = \alpha_{\text{nominal}}^{-1} \overline{AB}$$

$$= \pi \left(\frac{d_{\text{nozzle}}}{2}\right)^2 \sqrt{(x_2 - x_1)^2 + (y_2 - y_1)^2}, \qquad (1)$$

where A_{nozzle} is the area of the nozzle and $\alpha_{\text{nominal}} = A_{\text{nozzle}}^{-1}$ and $\overline{AB}$ is the distance between the points *A* and *B* (*see* **Note 2**).

G1: the machine rapidly moves the printing head to the specified point. This command is followed by the same options as *G1*. It is often used to move the printing head when the machine is not extruding.

A snippet of G-code is reported in Table 1.

Table 1
Snippet of G-code

1	G10
2	G0 F12000.00 X60.00000 Y50.00000 Z0.60000
3	G0 F1000.00000 X60.00000 Y50.00000
4	G11
5	G1 X60.00000 Y175.00000 Z0.60000 E238.762456

3.2 Printing of Functional Graded Structures

The fabrication of filaments with variable cross section is achieved by controlling the rate at which the material is extruded. In practice this is obtained by tailoring the extrusion parameter E in the G-code. When the parameter E is calculated as in Eq. 1, the diameter of the printed filament d_{filament} is equal to that of the nozzle d_{nozzle}. By contrast, if d_{nozzle} in Eq. 1 is substituted with a value $d < d_{\text{nozzle}}$, then $d_{\text{filament}} < d_{\text{nozzle}}$. Similarly, if d_{nozzle} in Eq. 1 is substituted with a value $d > d_{\text{nozzle}}$, then $d_{\text{filament}} > d_{\text{nozzle}}$. The approach presented here exploits the possibility of *under*-extruding and *over*-extruding the material through the nozzle with respect to the nominal demand. Over-extrusion results in a filament with a greater diameter than the nozzle size, while the opposite is true for under-extrusion. The ability to control E point to point allows us to extrude filaments with continuously variable diameter.

The diameter of the filament and that of the nozzle will be different when we under-extrude or over-extrude material. The expected diameter of the extruded filament can be derived from Eq. 1. If $\alpha \neq \alpha_{\text{nominal}}$ in Eq. 1, we obtain

$$d_{\text{filament}} = 2(\pi\alpha)^{-1/2}, \tag{2}$$

which is based purely on incompressible flow continuity considerations. Equation 2 relates a chosen value of α—the modified process parameter, with the expected diameter of the extruded filament d_{filament} during printing.

Firstly, calibration of the diameter of the printed filament should be performed and compared with that predicted by the incompressible continuity-based assumption of over-extrusion leading to a diameter larger than the nozzle diameter.

1. Using printing nozzles of different sizes, print long fibers (Fig. 1), each with a different value of α.

2. Measure filament diameter at randomly chosen five locations using an optical microscope equipped with a camera.

3. Plot the measured diameter values against the process parameter α (Fig. 2; *see* **Notes 3** and **4**).

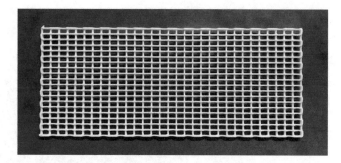

Fig. 1 The sample used to measure the diameter of filaments. The filaments running in the direction of the rectangle major side have been produced with different values of parameter α

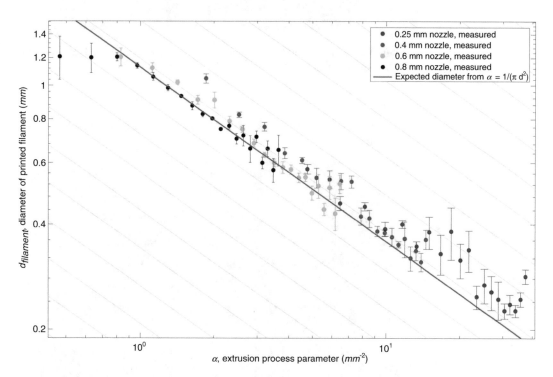

Fig. 2 Master calibration curve of the printed filament diameter as a function of extrusion process parameter (α). Each dot represents the mean ± standard deviation of five independent measurements. Four nozzle sizes have been used (0.25 mm, 0.4 mm, 0.6 mm, and 0.8 mm). The pink line is the expected values calculated using Eq. 2

3.3 Example Geometry and Mechanical Characterization

In the following example, details will be provided for the manufacturing of a rectangular specimen with functional graded features along its major side and for its mechanical characterization. The protocol shall be adapted for the manufacturing of complex-shaped objects.

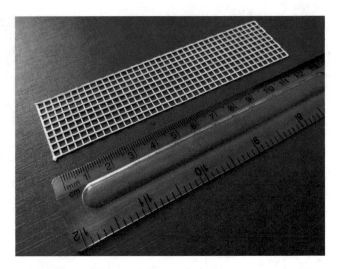

Fig. 3 Example of a specimen used for the buckling test. The graded filaments are those running along the major side

Table 2
G-code instructions for the fabrication of a functionally graded filament sample

G1 X160.00000 Y51.00000 Z0.64000 E4.218961; Segment #1
G1 X160.00000 Y52.00000 Z0.64000 E4.238660; Segment #2
G1 X160.00000 Y53.00000 Z0.64000 E4.258358; Segment #3
G1 X160.00000 Y54.00000 Z0.64000 E4.278056; Segment #4
G1 X160.00000 Y55.00000 Z0.64000 E4.297755; Segment #5
G1 X160.00000 Y56.00000 Z0.64000 E4.317453; Segment #6
G1 X160.00000 Y57.00000 Z0.64000 E4.337152; Segment #7
G1 X160.00000 Y58.00000 Z0.64000 E4.356850; Segment #8
G1 X160.00000 Y59.00000 Z0.64000 E4.376548; Segment #9
G1 X160.00000 Y60.00000 Z0.64000 E4.396247; Segment #10

1. Produce a rectangular sample made of two layers, as shown in Fig. 3. Each filament running along the major side is divided into 20 segments that is printed using a decreasing value of the extrusion rate. In the sample reported in Fig. 3, the filament diameter d_{filament} varies between 0.42 mm and 0.6 mm, given a space between filaments of 3.2 mm. The G-code for the manufacturing of a functionally graded filament is given in Table 2. The detailed G-code for the production of the

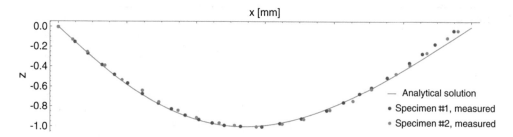

Fig. 4 (a) Test setup and buckled film after compression. **(b)** Comparison between the analytical model and the experimental tests for two films with graded properties

rectangular functionally graded specimen is reported in the Appendix.

2. The performance of functionally graded sample can be tested under buckling. If a thin strip or film with variable stiffness is subjected to buckling, the buckled shape is asymmetric. Place the specimen between two flat plates of a tensile test machine (Fig. 4a), and bring the specimen to a compression level of the critical buckling load.

3. Acquire a picture of the buckled profile, and extract the deformed shape using a software tool such as WebPlotDigitizer [22] (Fig. 4a).

   (a) The experimentally observed buckled shape can now be validated using the following analysis developed for thin films with graded properties. The buckled shape is theoretically obtained by solving the differential equation for the buckling of a Euler column with variable cross section:

$$\frac{d^2}{dx^2}\left[EI(x)\frac{d^2w}{dx^2}\right] + P\frac{d^2w}{dx^2} = 0. \tag{3}$$

   Exact solution for this differential equation can be found in [22]. For the thin films analyzed here, $EI(x) = E\frac{\pi}{4}r(x)^4$ is the variable bending stiffness. If we assume that the radius of the filaments varies linearly as $r(x) = Ax + B$, the buckled shape is given by

$$w(x) = \frac{\sqrt{2}\,bL(bx-1)\sec\left(\frac{\pi}{bL}\right)\sin\left(\frac{(bL-1)\pi x}{L(bx-1)}\right)}{\pi\sqrt{L|b| - bL^2|b|}}, \tag{4}$$

   where $b = -A/B$ and L is the length of the film (Fig. 4b).

4   Notes

1. Process can be adapted for use with other materials and/or 3D printers.

2. Over- or under-extrusion can be implemented on other 3D printing machines. However, different calculation for the feeding parameter might be necessary depending upon the hardware. For example, Ultimaker 3 printer (Ultimaker) makes a different usage of the extrusion parameter E within the G-code: instead of the volume of the extruded material, as for the Ultimaker 2, the value of E represents the length of the filament fed into the machine. The G-code for this printer would look exactly like the one shown in the previous section; only the numerical value for the parameter E would be different. For example, we assume that the spool of material is made of a filament of diameter d_{spool} and that the diameter of the nozzle is d_{nozzle}. Within a movement between two points at a distance l from each other, it is possible to equate the volume of material before and after the extrusion. Using L as the length of filament fed into the printer from the back, we obtain

$$l\pi \left(\frac{d_{\text{nozzle}}}{2}\right)^2 = L\pi \left(\frac{d_{\text{spool}}}{2}\right)^2.$$

By definition $E = L$. The value of E correspondent to the nozzle size is therefore given by

$$E_{\text{nominal}} = L_{\text{nominal}} = l \left(\frac{d_{\text{nozzle}}}{d_{\text{spool}}}\right)^2 = l\,\beta_{\text{nominal}}.$$

Thus, the parameter that controls the diameter of the printed filament in the Ultimaker 3 printer machine is

$$\beta = \left(\frac{d_{\text{filament}}}{d_{\text{spool}}}\right)^2,$$

from which we can calculate the expected printed diameter as

$$d_{\text{filament}} = d_{\text{spool}} \sqrt{\beta}.$$

3. From the graph plotted in Fig. 2, the diameter that is possible to extrude is in the range $[d_{\text{nozzle}}, 2d_{\text{nozzle}}]$. It is suggested to avoid filaments with a diameter smaller than the nominal diameter of the nozzle as this would increase the variability in the dimension of the printed filament.

4. The master calibration curve, such as the one presented in Fig. 2, must be constructed for each machine and each nozzle size to assess the accuracy of the process and identify the upper and lower limits achievable.

Acknowledgment

This work has received funding from the EPSRC Doctoral Prize Fellowship (Alessandra Bonfanti) and the HORIZON2020 innovation program under the Marie Sklodowska-Curie (Grant Agreement No. 643050, Loris Domenicale). Thanks are due to Dr. Chris Lovell for valuable comments on an earlier draft.

Appendix: G-Code to Print 2 × 2 Sample Using Ultimaker 2

```
;START_OF_HEADER
;HEADER_VERSION:0.1
;FLAVOR:Griffin
;GENERATOR.NAME:Cura_SteamEngine
;GENERATOR.VERSION:2.7.0
;GENERATOR.BUILD_DATE:2017-08-30
;TARGET_MACHINE.NAME:Ultimaker 3 Extended
;EXTRUDER_TRAIN.0.INITIAL_TEMPERATURE:210
;EXTRUDER_TRAIN.0.MATERIAL.VOLUME_USED:49590
;EXTRUDER_TRAIN.0.MATERIAL.GUID:506c9f0d-e3aa-4bd4-b2d2-
23e2425b1aa9
;EXTRUDER_TRAIN.0.NOZZLE.DIAMETER:0.4
;EXTRUDER_TRAIN.0.NOZZLE.NAME:AA 0.4
;BUILD_PLATE.INITIAL_TEMPERATURE:60
;PRINT.TIME:60
;PRINT.SIZE.MIN.X:9
;PRINT.SIZE.MIN.Y:6
;PRINT.SIZE.MIN.Z:0.27
;PRINT.SIZE.MAX.X:179.675
;PRINT.SIZE.MAX.Y:184.003
;PRINT.SIZE.MAX.Z:8.07
;END_OF_HEADER
;G-code generated with a MATLAB script

T0
G92 E0

M109 S210
G0 F15000 X9 Y6 Z2
G280
G1 F1500 E-6.5;LAYER_COUNT:2
;LAYER:0
M107
```

```
M204 S500
M205 X6 Y6
G0 F9000 X20.00000 Y20.00000 Z0.32000

;Starting extrusion
G1 F1000.00000
G1 X160.00000 Y50.00000 Z0.32000 E2.820377
;Layer 1
G1 X180.00000 Y50.00000 Z0.32000 E3.214345
G1 X180.00000 Y55.00000 Z0.32000 E3.312836
G1 X160.00000 Y55.00000 Z0.32000 E3.706804
G1 X160.00000 Y60.00000 Z0.32000 E3.805296
G1 X180.00000 Y60.00000 Z0.32000 E4.199263
;Layer 2
G10
G0 F12000.00 X160.00000 Y50.00000 Z0.64000
G0 F1000.00000 X160.00000 Y50.00000
G11
G1 X160.00000 Y51.00000 Z0.64000 E4.218961 ;Segment #1
G1 X160.00000 Y52.00000 Z0.64000 E4.238660 ;Segment #2
G1 X160.00000 Y53.00000 Z0.64000 E4.258358 ;Segment #3
G1 X160.00000 Y54.00000 Z0.64000 E4.278056 ;Segment #4
G1 X160.00000 Y55.00000 Z0.64000 E4.297755 ;Segment #5
G1 X160.00000 Y56.00000 Z0.64000 E4.317453 ;Segment #6
G1 X160.00000 Y57.00000 Z0.64000 E4.337152 ;Segment #7
G1 X160.00000 Y58.00000 Z0.64000 E4.356850 ;Segment #8
G1 X160.00000 Y59.00000 Z0.64000 E4.376548 ;Segment #9
G1 X160.00000 Y60.00000 Z0.64000 E4.396247 ;Segment #10
G1 X170.00000 Y60.00000 Z0.64000 E4.593230
G1 X170.00000 Y59.00000 Z0.64000 E4.612929 ;Segment #1
G1 X170.00000 Y58.00000 Z0.64000 E4.632627 ;Segment #2
G1 X170.00000 Y57.00000 Z0.64000 E4.652325 ;Segment #3
G1 X170.00000 Y56.00000 Z0.64000 E4.672024 ;Segment #4
G1 X170.00000 Y55.00000 Z0.64000 E4.691722 ;Segment #5
G1 X170.00000 Y54.00000 Z0.64000 E4.711421 ;Segment #6
G1 X170.00000 Y53.00000 Z0.64000 E4.731119 ;Segment #7
G1 X170.00000 Y52.00000 Z0.64000 E4.750817 ;Segment #8
G1 X170.00000 Y51.00000 Z0.64000 E4.770516 ;Segment #9
G1 X170.00000 Y50.00000 Z0.64000 E4.790214 ;Segment #10
G1 X180.00000 Y50.00000 Z0.64000 E4.987198
G1 X180.00000 Y51.00000 Z0.64000 E5.006896 ;Segment #1
G1 X180.00000 Y52.00000 Z0.64000 E5.026594 ;Segment #2
G1 X180.00000 Y53.00000 Z0.64000 E5.046293 ;Segment #3
G1 X180.00000 Y54.00000 Z0.64000 E5.065991 ;Segment #4
G1 X180.00000 Y55.00000 Z0.64000 E5.085690 ;Segment #5
G1 X180.00000 Y56.00000 Z0.64000 E5.105388 ;Segment #6
G1 X180.00000 Y57.00000 Z0.64000 E5.125086 ;Segment #7
G1 X180.00000 Y58.00000 Z0.64000 E5.144785 ;Segment #8
```

```
G1 X180.00000 Y59.00000 Z0.64000 E5.164483 ;Segment #9
G1 X180.00000 Y60.00000 Z0.64000 E5.184181 ;Segment #10
M107

M104 S0
M104 T1 S0
;End of Gcode
```

Below is the sample generated with the code reported in the Appendix.

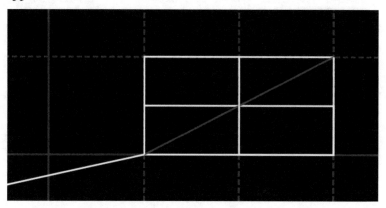

References

1. Ryan G, Pandit A, Apatsidis DP (2006) Fabrication methods of porous metals for use in orthopaedic applications. Biomaterials 27 (13):2651–2670

2. Bonfanti A (2016) Mechanics of structured materials and their biomedical applications. Doctoral dissertation, University of Southampton

3. Alessandra B, Stavros S, Atul B (2020) Structural analysis of cyclically periodic rings and its application to the mechanics of balloon expandable stents. International Journal of Solids and Structures (185–186):46–56

4. O'Brien FJ (2011) Biomaterials & scaffolds for tissue engineering. Mater Today 14(3):88–95

5. Do AV, Khorsand B, Geary SM, Salem AK (2015) 3D printing of scaffolds for tissue regeneration applications. Adv Healthc Mater 4(12):1742–1762

6. Gibson LJ, Ashby MF, Harley BA (2010) Cellular materials in nature and medicine. Cambridge University Press, Cambridge

7. Murphy CM, Haugh MG, O'Brien FJ (2010) The effect of mean pore size on cell attachment, proliferation and migration in collagen–glycosaminoglycan scaffolds for bone tissue engineering. Biomaterials 31 (3):461–466

8. Zhang Q, Lu H, Kawazoe N, Chen G (2014) Pore size effect of collagen scaffolds on cartilage regeneration. Acta Biomater 10 (5):2005–2013

9. Kasten P, Beyen I, Niemeyer P, Luginbühl R, Bohner M, Richter W (2008) Porosity and pore size of β-tricalcium phosphate scaffold can influence protein production and osteogenic differentiation of human mesenchymal stem cells: an in vitro and in vivo study. Acta Biomater 4(6):1904–1915

10. Di Luca A, Ostrowska B, Lorenzo-Moldero I, Lepedda A, Swieszkowski W, Van Blitterswijk C, Moroni L (2016) Gradients in pore size enhance the osteogenic differentiation of human mesenchymal stromal cells in three-dimensional scaffolds. Sci Rep 6:22898

11. Sobral JM, Caridade SG, Sousa RA, Mano JF, Reis RL (2011) Three-dimensional plotted scaffolds with controlled pore size gradients: effect of scaffold geometry on mechanical performance and cell seeding efficiency. Acta Biomater 7(3):1009–1018

12. Singh M, Dormer N, Salash JR, Christian JM, Moore DS, Berkland C, Detamore MS (2010) Three-dimensional macroscopic scaffolds with a gradient in stiffness for functional regeneration of interfacial tissues. J Biomed Mater Res A 94(3):870–876

13. Pompe W, Worch H, Epple M, Friess W, Gelinsky M, Greil P et al (2003) Functionally graded materials for biomedical applications. Mater Sci Eng A 362(1–2):40–60

14. Paul JP (1999) Strength requirements for internal and external prostheses. J Biomech 32(4):381–393

15. Huiskes R, Weinans H, Van Rietbergen B (1992) The relationship between stress shielding and bone resorption around total hip stems and the effects of flexible materials. Clin Orthop Relat Res 274:124–134

16. Yang S, Leong KF, Du Z, Chua CK (2001) The design of scaffolds for use in tissue engineering. Part I. Traditional factors. Tissue Eng 7(6):679–689

17. An J, Teoh JEM, Suntornnond R, Chua CK (2015) Design and 3D printing of scaffolds and tissues. Engineering 1(2):261–268

18. Rashed MG, Ashraf M, Mines RAW, Hazell PJ (2016) Metallic microlattice materials: a current state of the art on manufacturing, mechanical properties and applications. Mater Des 95:518–533

19. Woodfield TBF, Blitterswijk CV, Wijn JD, Sims TJ, Hollander AP, Riesle J (2005) Polymer scaffolds fabricated with pore-size gradients as a model for studying the zonal organization within tissue-engineered cartilage constructs. Tissue Eng 11(9–10):1297–1311

20. Cheon Deok J (2016) 3D Printer having variable nozzle head, and operation method therefor. WO2016072549 (Patent number)

21. WebPlotDigitizer - Extract data from plots, images, and maps. https://automeris.io/WebPlotDigitizer/. Accessed 21 Nov 2017

22. Wang CM, Wang CY (2004) Vol. 6. In: Exact solutions for buckling of structural members. CRC Press, Boca Raton

Part II

Biomaterials for Computer-Aided Tissue Engineering

Chapter 4

Photocurable Biopolymers for Coaxial Bioprinting

Marco Costantini, Andrea Barbetta, Wojciech Swieszkowski, Dror Seliktar, Cesare Gargioli, and Alberto Rainer

Abstract

Thanks to their unique advantages, additive manufacturing technologies are revolutionizing almost all sectors of the industrial and academic worlds, including tissue engineering and regenerative medicine. In particular, 3D bioprinting is rapidly emerging as a first-choice approach for the fabrication—in one step—of advanced cell-laden hydrogel constructs to be used for in vitro and in vivo studies. This technique consists in the precise deposition *layer-by-layer* of sub-millimetric hydrogel strands in which living cells are embedded. A key factor of this process consists in the proper formulation of the hydrogel precursor solution, the so-called bioink. Ideal bioinks should be able, on the one side, to support cell growth and differentiation and, on the other, to allow the high-resolution deposition of cell-laden hydrogel strands. The latter feature requires the extruded solution to instantaneously undergo a sol-gel transition to avoid its collapse after deposition.

To address this challenge, researchers are recently focusing their attention on the synthesis of several derivatives of natural biopolymers to enhance their printability. Here, we present an approach for the synthesis of photocurable derivatives of natural biopolymers—namely, gelatin methacrylate, hyaluronic acid methacrylate, chondroitin sulfate methacrylate, and PEGylated fibrinogen—that can be used to formulate tailored innovative bioinks for coaxial-based 3D bioprinting applications.

Key words Coaxial bioprinting, Photocurable polymers, Alginate, Bioink formulation

1 Introduction

Nowadays, additive manufacturing systems represent a fast and cost-effective biofabrication technology, able to create 3D objects with high precision, high resolution, and high reproducibility [1]. Thanks to these attractive features, additive manufacturing and bioprinting are rapidly becoming a first-choice approach for the production of engineered materials for tissue engineering (TE) [2]. In particular, 3D bioprinting represents one of the most innovative and promising approaches for the fabrication in one step of advanced cell-laden constructs, which aim to recapitulate the complexity of human organs and tissues. Like other additive

Alberto Rainer and Lorenzo Moroni (eds.), *Computer-Aided Tissue Engineering: Methods and Protocols*,
Methods in Molecular Biology, vol. 2147, https://doi.org/10.1007/978-1-0716-0611-7_4,
© Springer Science+Business Media, LLC, part of Springer Nature 2021

manufacturing techniques, 3D bioprinting allows the free-form fabrication of 3D shapes through the deposition of consecutive layers of hydrogel strands that contain living cells [3].

To succeed in the biofabrication process, a key factor consists in formulating a proper hydrogel precursor solution, the *bioink* [4]. In fact, an ideal bioink should possess two fundamental features: first, the bioink should instantaneously turn into a gel as soon as it is deposited; second, it should support cell migration, proliferation, and differentiation. Hence, formulating a bioink is an extremely complex task as most natural biopolymers in their pristine version (i.e., not chemically modified) do not undergo a fast sol-gel transition under mild and cell-friendly conditions.

To overcome this issue, a common strategy consists in tuning the rheological properties of bioinks to have a pronounced shear-thinning behavior [5–7]. In this way, bioinks can easily flow (liquid-like behavior) under applied pressure and rapidly form a gel (solid-like behavior) after extrusion when the force is removed. However, this strategy is generally time-consuming and requires a thorough rheological characterization and a fine optimization of all bioprinting parameters.

Another approach consists in blending one or more biopolymers with an additional templating agent [8, 9]. The main role of such compound consists in improving the processability and printability of other biopolymers in blend, without affecting cell viability.

So far, the most common templating agent used in 3D bioprinting experiments is alginate. Alginate is a polysaccharide extracted from brown algae that is capable of undergoing a fast, reversible, and cell-friendly gelation in presence of divalent ions (such as Ca^{2+}, Sr^{2+}, Ba^{2+}). This unique feature generally allows the fabrication of high-resolution and high-shape fidelity constructs, with a cell viability above 80%. Additionally, alginate is commercially available in a wide range of molecular weight values ranging from few tens to hundreds of kDa, allowing researchers to easily tune the rheological properties of the final bioinks.

Alginate-based bioinks have been deposited in 3D using several strategies, including inkjet printing [10, 11], divalent ion spraying [12, 13], printing in a coagulation bath [14, 15], pre-crosslinking [16], and coaxial nozzle extrusion [17, 18]. Among these, coaxial nozzle extrusion represents one of the most versatile and powerful approaches. This approach consists of two coaxially mounted nozzles in which an alginate-based bioink and a calcium chloride solution flow. Hydrogel fiber gelation takes place instantaneously at the tip of the coaxial extruder where the two solutions meet, allowing the fine deposition of hydrogel strands [19].

A unique advantage of coaxial systems consists in the possibility to deposit either bulky or hollow (i.e., perfusable) tiny fibers down to ~100 μm in diameter. This can be easily achieved by supplying the alginate-based bioink either through the internal (bulky fibers) or the external (hollow fibers) nozzle [20, 21].

Due to its reversible crosslinking nature, alginate-based bioinks are generally blended with photocurable biopolymers—such as gelatin methacrylate (GelMA), hyaluronic acid methacrylate (HAMA), chondroitin sulfate 2-aminoethyl methacrylate (CS-AEMA) [17], PEGylated(monoacrylate) fibrinogen [18], etc.—that can be used to fabricate chemically crosslinked and stable constructs after a short exposure to mild UV radiation (typically at 365 nm).

In the next sections, we will describe how to formulate tailored alginate-based bioinks to engineer cartilage and skeletal muscle tissue in vitro.

2 Materials

The synthesis of photo-crosslinkable derivatives of natural polymers should be performed in a wet lab equipped with fume hoods and a freeze drier.

2.1 Synthesis of Gelatin Methacrylate (GelMA) and Hyaluronic Acid Methacrylate (HAMA)

1. Gelatin (type A, 300 bloom) or hyaluronic acid (*see* **Note 1**).
2. Methacrylic anhydride (MA).
3. Round-bottom flasks.
4. Phosphate buffer saline (PBS).
5. Dialysis tubing (MWCO = 2 kDa).
6. Deionized water (DIW).

2.2 Synthesis of Chondroitin Sulfate 2-Aminoethyl Methacrylate (CS-AEMA)

1. Chondroitin sulfate.
2. 2-Aminoethyl methacrylate (AEMA).
3. 1-Ethyl-3-(3-dimethylaminopropyl)carbodiimide hydrochloride (EDC).
4. N-Hydroxysuccinimide (NHS).
5. MES buffer: 50 mM 2-(N-morpholino)ethanesulfonic acid, pH = 6.5.
6. NaCl.
7. Deionized water (DIW).
8. Dialysis tubing (MWCO = 2 kDa).
9. Round-bottom flasks.

2.3 Synthesis of PEG(Monoacrylate)-Fibrinogen

1. Fibrinogen (recombinant human).
2. Phosphate buffer: 50 mM PBS, pH 7.4.
3. PBS/urea solution: 50 mM PBS, 8 M urea.

4. PEG-diacrylate solution: 250–300 mg/mL PEG-DA (4–20 kDa; *see* **Note 2**) in PBS/urea buffer. Freshly prepared.

5. Tris(2-carboxyethyl)phosphine hydrochloride (TCEP-HCl).

6. Acetone.

7. Dialysis tubing (MWCO = 12–14 kDa).

8. D-Glucose.

2.4 Preparation of the Bioink

1. Heated magnetic stirrer.

2. Screw top glass vial, preferably of brown glass (*see* **Note 3**).

3. Low molecular weight alginate (FMC Biopolymers).

4. GelMA, HAMA, CS-AEMA, or PEG-fibrinogen (*see* **Note 4**).

5. 2-Hydroxy-4′-(2-hydroxyethoxy)-2-methylpropiophenone photoinitiator (Irgacure 2959, BASF).

6. HEPES/FBS buffer: 25 mM HEPES buffer containing 10% v/v fetal bovine serum (FBS).

7. Sterile tubes (1.5–2 mL).

8. 0.2 μm syringe filter.

9. 1 mL syringe with 19G needle.

10. Living cells (according to the application).

11. Cell culture plasticware, media, and consumables.

2.5 Bioprinting Experiment

1. 3D bioprinter (or three-axis controlled motorized stage).

2. Coaxial needle (*see* **Note 5**).

3. Two syringe pumps (*see* **Note 6**).

4. UV light source at 365 nm.

5. ~700 μL of sterile bioink (*see* Subheading 2.4).

6. Crosslinking solution: sterile 300 mM aq. calcium chloride solution.

7. Two gas-tight glass syringes (1 mL), sterilized by autoclaving.

8. PTFE tubing to connect the syringes to the coaxial needle, sterilized by autoclaving.

9. Sterile spatula.

10. Sterile plasticware.

11. Cell culture media.

2.6 Immuno-fluorescence Evaluation of Structural and Functional Proteins in 3D Bioprinted Constructs

1. Fixation buffer: 2% paraformaldehyde in PBS.

2. Permeabilization solution: 0.2% v/v Triton X-100 in PBS.

3. Blocking solution: 0.1% v/v Triton X-100, 1% w/v glycine, 10% normal goat serum in PBS.

4. Washing solution: 0.2% v/v Triton X-100 and 1% w/v bovine serum albumin in PBS.

5. Primary antibodies.

6. Fluorophore-conjugated secondary antibodies.

7. DAPI nuclear stain (1:10000).

8. Widefield or confocal fluorescence microscope.

3   Methods

3.1   Synthesis of Gelatin Methacrylate (GelMA) and Hyaluronic Acid Methacrylate (HAMA)

1. Dissolve 1 g of gelatin in 10 mL of PBS under magnetic stirring at 50 °C.

2. Dropwise add 0.8 mL of MA under vigorous stirring. The reaction mixture must be kept at 50 °C.

3. Stop the reaction after 2 h.

3.1.1   Gelatin Methacrylate

4. Dilute the reaction mixture to 100 mL, transfer it into dialysis tubes, and dialyze at 40 °C against distilled water for 3 days with dialysate changes three times per day.

5. Transfer the solution into round-bottom flasks and freeze-dry.

3.1.2   Hyaluronic Methacrylate

1. Dissolve 1 g of HA in 100 mL of DIW under magnetic stirring at room temperature.

2. Dropwise add 20-fold excess of MA vs. the disaccharide repeating unit (2-acetamido-2-deoxy-α-D-glucose and β-D-glucuronic acid; residue MW, 401). For a solution containing 1 g of HA, this corresponds to a volume of 0.37 mL.

3. During MA addition, continuously adjust the pH of the reaction mixture to 8.0 by adding aliquots of 5 M NaOH solution.

4. Leave the reaction mixture under stirring on ice for 24 h.

5. Perform dialysis at room temperature for 3 days with dialysate changes three times per day, followed by freeze drying.

3.1.3   Synthesis of Chondroitin Sulfate 2-Aminoethyl Methacrylate (CS-AEMA)

1. Dissolve 1 g of chondroitin sulfate under magnetic stirring at room temperature in 100 mL of MES buffer containing 0.5 M of NaCl.

3. Sequentially add 0.27 g of NHS and 0.85 g EDC and, after ca. 5 min, 0.38 g AEMA.

3. Leave the reaction mixture under stirring for 24 h at room temperature.

4. Follow **step 5** of Subheading 3.1.2.

3.1.4 Synthesis of PEG(monoacrylate)-Fibrinogen

1. Prepare 100 mL of a 7 mg/mL solution of fibrinogen in PBS/urea buffer.

2. Add TCEP-HCl at a molar ratio 68:1 with respect to fibrinogen cysteines, and leave the solution under stirring for 15 min at 25 °C until complete dissolution.

3. Add the PEG-DA solution so that the PEG-fibrinogen molar ratio is 145:1.

4. Leave the reaction overnight in the dark at 25 °C.

5. Precipitate the final PEGylated fibrinogen product in fivefold excess acetone at room temperature for 20 min under stirring.

6. Centrifuge at $4000 \times g$ for 20 min.

7. Redissolve the resulting pellet to 20 mg/mL protein concentration in PBS/urea buffer.

8. Dialyze the PEGylated fibrinogen against PBS at 4 °C for 2 days with twice-daily changes.

9. Lyophilize and store the PEGylated product in argon at -80 °C.

3.2 Preparation of the Bioink and Cell Resuspension

3.2.1 Preparation of the Bioink

1. Weigh 40 mg of alginate (*see* **Note 7**) in a 2 mL tube.

2. Weigh the photocurable biopolymer(s) to be added to the bioink (*see* **Note 8**).

3. Add 1 mL of HEPES/FBS buffer (*see* **Note 9**).

4. Vortex until dissolution.

5. Centrifuge to remove air bubbles.

6. Filter the bioink using 0.22 μm syringe filters in a laminar flow hood.

3.2.2 Resuspension of the Cells Within the Bioink

The following instructions must be conducted in a laminar flow hood.

1. Following standard procedures, detach adherent cells, resuspend them in suitable culture medium, and calculate cell concentration.

2. Place the volume of cell suspension containing the desired number of cells in a sterile 2 mL tube, and centrifuge for 5 min at $180 \times g$ (*see* **Note 10**).

3. Gently remove the supernatant (*see* **Note 11**).

4. Add 1 mL of sterile-filtered bioink solution and gently resuspend the cells.

3.3 Bioprinting Experiment

Bioprinting should be performed within a biological safety hood.

1. Create a G-code for the desired construct geometry (*see* **Note 12**).

2. Connect the coaxial needle to the *Z*-axis of the bioprinter/ three-axis coordinated motion control system.

3. Calibrate the *Z*-axis (i.e., set the zero level of the axis on top of the surface you will use to print).

4. Fill a 1 mL syringe with the bioink.

5. Fill another 1 mL syringe with the crosslinking solution.

6. Mount the two syringes on the syringe pumps and connect them to the inner/outer needles of the coaxial assembly using PTFE tubing.

7. Start the flow of the two pumps.

8. Start the G-code.

9. After bioprinting, collect your deposited sample with a sterile spatula and place in a Petri dish.

10. Perform photo-crosslinking under UV light (*see* **Note 13**).

11. After UV crosslinking, take your sample and place in a multi-well plate and add the desired cell culture medium.

3.4 Immuno-fluorescence Evaluation of Structural and Functional Proteins in 3D Bioprinted Constructs

The following protocol for immunofluorescence staining has been established for the processing of bioprinted constructs and can be further optimized according to the specific application.

1. Wash constructs in PBS 5 min at RT.

2. Fix constructs in fixation buffer for 10 min at 4 °C (*see* **Note 14**).

3. Wash constructs in PBS for PFA excess removal 5 min at RT.

4. Permeabilize in permeabilizing solution for 1 h at RT.

5. Incubate in blocking solution for 1 h at RT.

6. Incubate with primary antibodies (dilution according to the manufacturer's instructions) in blocking solution (1 h at RT or o/n at 4 °C).

7. Wash constructs in washing solution twice for 15 and 5 min at RT.

8. Incubate with secondary antibodies (dilution according to the manufacturer's instructions) in blocking solution for 1 h at RT.

9. Wash constructs twice in washing solution for 10 min and once in PBS for 10 min at RT.

10. Incubate with DAPI (1:10000) for 10 min at RT to label nuclei.

11. Mount constructs with Aqua-Poly/Mount (Polysciences) mounting medium.

12. Specimens can be observed under a widefield or confocal fluorescence microscope.

4 Notes

1. Gelatin (type A or type B) from any source and with any molecular weight can be used. Similarly, the methacrylation reaction can be performed with any type of hyaluronic acid, independently of its molecular weight.

2. For the synthesis of photocurable PEGylated fibrinogen, PEG-diacrylate (PEGDA) at different molecular weights can be used. Typically, 4, 6 or 20 kDa PEGDA are used in the synthesis. Such polymers can be prepared by reacting PEG with acryloyl chloride in anhydrous conditions (for more details *see* Cruise et al. [22]) or bought from local suppliers.

3. The bioink solution can be prepared in any container, but care must be taken to protect the solution from light after photo-initiator addition.

4. Bioink formulation can vary from one experiment to another. In particular, its formulation is generally influenced by the cell type used in the experiment. Therefore, in order to provide the best microenvironment to the cells for their growth, proliferation, and differentiation, bioinks can contain a single photo-curable biopolymer or a combination of them.

5. We recommend using a 19G outer needle and a 25G inner needle. While coaxial needle assemblies can be bought from specialized providers, lab-built solutions can easily be assembled starting from conventional blunt tip needles.

6. Two pumps are always needed to supply the bioink and the crosslinking solution ($CaCl_2$), respectively, to the inner and outer needle of the coaxial system. Additional pumps may be required to supply multiple bioinks simultaneously.

7. All the instructions are meant for 1 mL of bioink solution, but the quantities can be scaled up for larger volumes. For small volumes (i.e., less than 2 mL), we recommend using 1.5 mL or 2 mL tubes. For larger volumes (> 2 mL), we recommend using screw-cap glass vials.

8. We suggest the following concentration ranges: GelMA 1÷10% w/w, HAMA 1÷3% w/w, CS-AEMA 2÷10% w/w, and PEG-fibrinogen 0.8÷1.5% w/w.

9. Bioink buffer can be also formulated without FBS.

10. Cell density may vary according to the cell type and the application. Typically, cell density values in the range $5÷50 \times 10^6$ cells/mL are recommended. However, a large cell number can lead to difficulties in cell resuspension and formation of cell aggregates.

11. We recommend to carefully aspirate as much supernatant as possible, to avoid diluting the bioink.

12. We recommend the following parameters to achieve the best printing results: layer thickness = 100 μm, printing speed = 235 mm/min, and fiber-to-fiber distance = 400 μm.

13. According to the UV light intensity generated by the source (generally a UV lamp), the crosslinking time may change. A low-dose UV irradiation of 1.3 mW/cm^2 for 5 min at 365 nm should be enough to crosslink the hydrogel while still not affecting cell viability.

14. Alternatively, fixation can be performed in ice cold methanol for 1 h at 4 °C.

References

1. Gibson I, Rosen D, Stucker B (2015) Additive manufacturing technologies. Springer, New York, NY

2. Murphy SV, Atala A (2014) 3D bioprinting of tissues and organs. Nat Biotechnol 32:773

3. Kang HW, Lee SJ, Ko IK, Kengla C, Yoo JJ, Atala A (2016) A 3D bioprinting system to produce human-scale tissue constructs with structural integrity. Nat Biotechnol 34:312–319

4. Ji S, Guvendiren M (2015) Recent advances in bioink design for 3D bioprinting of tissues and organs. Front Bioeng Biotechnol 5:23–31

5. Hölzl K, Lin S, Tytgat L, Van Vlierberghe S, Gu L, Ovsianikov A (2016) Bioink properties before, during and after 3D bioprinting. Biofabrication 8:32002

6. Chimene D, Lennox KK, Kaunas RR, Gaharwar AK (2016) Advanced bioinks for 3D printing: a materials science perspective. Ann Biomed Eng 44:2090–2102

7. Panwar A, Tan L (2016) Current status of bioinks for micro-extrusion-based 3D bioprinting. Molecules 21:E685

8. Armstrong JPK, Burke M, Carter BM, Davis SA, Perriman AW (2016) 3D bioprinting using a templated porous bioink. Adv Healthc Mater 5:1724

9. Axpe E, Oyen ML (2016) Applications of alginate-based bioinks in 3D bioprinting. Int J Mol Sci 17:1976

10. Xu T, Zhao W, Zhu JM, Albanna MZ, Yoo JJ, Atala A (2013) Complex heterogeneous tissue constructs containing multiple cell types prepared by inkjet printing technology. Biomaterials 34:130–139

11. Xu T, Baicu C, Aho M, Zile M, Boland T (2009) Fabrication and characterization of bio-engineered cardiac pseudo tissues. Biofabrication 1:35001

12. Kosik-Kozioł A, Costantini M, Bolek T, Szoke K, Barbetta A, Brinchmann JE, Święszkowski W (2017) PLA short sub-micron fibers reinforcement of 3D bioprinted alginate constructs for cartilage regeneration. Biofabrication 9:044105

13. Yeo MG, Lee JS, Chun W, Kim GH (2016) An innovative collagen-based cell-printing method for obtaining human adipose stem cell-laden structures consisting of core-sheath structures for tissue engineering. Biomacromolecules 17:1365–1375

14. Nishiyama Y, Nakamura M, Henmi C, Yamaguchi K, Mochizuki S, Nakagawa H, Takiura K (2007) Fabrication of 3D cell supporting structures with multi-materials using the bio-printer. In: ASME 2007 International manufacturing science and engineering conference, american society of mechanical engineers, New York City

15. Khalil S, Sun W (2009) Bioprinting endothelial cells with alginate for 3D tissue constructs. J Biomech Eng 131:111002

16. Diogo GS, Gaspar VM, Serra IR, Fradique R, Correia IJ (2014) Manufacture of β- TCP/alginate scaffolds through a Fab@home model for application in bone tissue engineering. Biofabrication 6:25001

17. Costantini M, Idaszek J, Szöke K, Jaroszewicz J, Dentini M, Barbetta A, Brinchmann JE, Święszkowski W (2016) 3D bioprinting of BM-MSCs-loaded ECM biomimetic hydrogels for in vitro neocartilage formation. Biofabrication 8:35002

18. Costantini M, Testa S, Mozetic P, Barbetta A, Fuoco C, Fornetti E, Tamiro F, Bernardini S, Jaroszewicz J, Święszkowski W, Trombetta M, Castagnoli L, Seliktar D, Garstecki P, Cesareni G, Cannata S, Rainer A, Gargioli C (2017) Biomaterials 131:98–110

19. Colosi C, Costantini M, Latini R, Ciccarelli S, Stampella A, Barbetta A, Massimi M, Devirgiliis LC, Dentini M (2014) Rapid prototyping of chitosan-coated alginate scaffolds through the use of a 3D fiber deposition technique. J Mater Chem B 2:6779–6791

20. Li Y, Liu Y, Jiang C, Li S, Liang G, Hu Q (2016) A reactor-like spinneret used in 3D printing alginate hollow fiber: a numerical study of morphological evolution. Soft Matter 12:2392–2399

21. Costantini M, Colosi C, Święszkowski W, Barbetta A (2018) Co-axial wet-spinning in 3D bioprinting: state of the art and future perspective of microfluidic integration. Biofabrication 11:012001

22. Cruise GM, Scharp DS, Hubbell JA (1998) Characterization of permeability and network structure of interfacially photopolymerized poly(ethylene glycol) diacrylate hydrogels. Biomaterials 14:1287

Chapter 5

Synthesis of an UV-Curable Divinyl-Fumarate Poly-ε-Caprolactone for Stereolithography Applications

Alfredo Ronca, Sara Ronca, Giuseppe Forte, and Luigi Ambrosio

Abstract

The limited number of commercially available photocrosslinkable resins for stereolithography has often been considered the main limitation of this technique. In this manuscript, a photocrosslinkable poly--ε-caprolactone (PCL) has been synthesized by a two-step method starting from ring opening polymerization (ROP) of ε-caprolactone. Hydroxyethyl vinyl ether (HEVE) has been used both as the initiator of ROP and as photo-curable functional group to obtain a vinyl poly-ε-caprolactone (VPCL). The following reaction of VPCL with fumaryl chloride (FuCl) results in a divinyl-fumarate polycaprolactone (VPCLF). Moreover, a catalyst based on Al, instead of the most popular Tin(II) 2-ethylhexanoate, has been employed to reduce the cytotoxicity of the material. VPCLF has been successfully used, in combination with N-vinyl-pyrrolidone (NVP), to fabricate 3D porous scaffolds by micro-stereolithography (μ-SL) with mathematically defined architectures.

Key words Tissue engineering, Stereolithography, Photocrosslinkable polymer, Polycaprolactone

1 Introduction

Rapid prototyping (RP) is a term which embraces a range of new technologies for producing a physical model directly from computer-aided design (CAD) data [1–3]. Unlike conventional machining, which involves constant removal of materials, RP builds parts by selectively adding material layer by layer, as specified by a computer program, where each layer represents the shape of the cross section of the model at a specific level. The many advantages of RP technologies include the following: parts can be easily customized and personalized, no need for special tooling for part fabrication, and material waste is greatly reduced [4, 5]. Also in designing surgical tools, implants, and other biomedical devices, these additive fabrication methods have been largely used in the last decade. Specially in the tissue engineering (TE) field that represents a new concept to treat problems associated with failing tissues and

Alberto Rainer and Lorenzo Moroni (eds.), *Computer-Aided Tissue Engineering: Methods and Protocols*,
Methods in Molecular Biology, vol. 2147, https://doi.org/10.1007/978-1-0716-0611-7_5,
© Springer Science+Business Media, LLC, part of Springer Nature 2021

organs, the combination of automation and flexibility in design makes RP very suitable for the generation of personalized implants [6–8]. Different processing techniques and methodologies have been proposed to optimize final scaffold performances, but in the last years, RP technologies have reached the best results in terms of external shape and size, surface morphology, and internal architecture [9, 10]. Among all the RP techniques, stereolithography (SLA) has become a valuable tool for the fabrication of biocompatible tissue engineering scaffolds, due to its ability to fabricate precise internal architectures and external geometries, which match those of human tissues [11, 12]. The working principle of SLA is based on spatially controlled solidification of a liquid photopolymerizable resin through a computer-controlled laser beam or a digital light projection [13]. While in most fabrication techniques the smallest details are 50–200 μm in size, many commercially available SLA setups can build objects that measure several cubic centimeters at an accuracy of 20 μm [12, 14, 15]. Most of the commercially available resins for SLA are based on low-molecular-weight acrylate polymers that are generally glassy, rigid, and brittle materials [16, 17]. The functionalization of resorbable polyesters, such as poly(ethylene glycol) (PEG), polylactide (PLA), and poly (ε-caprolactone) (PCL), with unsaturated groups and subsequent ultraviolet (UV) crosslinking has been extensively studied to overcome this limitation [16–18]. However, the diffusion of unpolymerized methacrylates is one of the most important factors causing irritation in tissues [19, 20]. To overcome these problems, here we report the synthesis of a trifunctional photocrosslinkable vinyl-fumarate PCL for tissue engineering applications. Vinyl-poly-ε-caprolactone has been synthesized by ring opening polymerization (ROP) using an alcohol/methylaluminum diphenolate system (AlMe(OR)$_2$) as a catalyst and 2-hydroxyethyl vinyl ether (HEVE) as initiator [21, 22]. Subsequently the vinyl-terminated PCL was reacted with fumaryl chloride in order to obtain a divinyl-fumarate PCL (VPCLF). Networks were formed by UV irradiation (365 nm) of VPCLF macromers using ethyl-2,4,6-trimethylbenzoylphenylphosphinate (Lucirin TPO-L) as a biocompatible initiator and N-vinyl-2-pyrrolidone (NVP) as a crosslinking agent. In this way, the VPCLF/NVP resin was used to realize porous scaffolds, based on triply periodic minimal surface (TPMS) geometries, by SLA technique.

2   Materials

The synthesis requires a standard glassware equipment for operation of air- and moisture-sensitive materials (Schlenk line, etc.). All the materials involved in the synthesis of PCL and subsequent reaction with fumaryl chloride should be anhydrous/dry and flushed with nitrogen to remove traces of oxygen and water as much as possible. All manipulations of both air- and moisture-sensitive compounds were performed under a protective atmosphere (nitrogen or argon) using a glove box.

2.1 Synthesis of VPCLF

1. Dichloromethane (DCM), anhydrous.
2. Trimethylaluminum in toluene.
3. 2-Hydroxyethyl vinyl ether (HEVE).
4. ε-CL (Aldrich). Further dried before use by 4 Å molecular sieves.
5. Isobutylaluminum bis(2,6-di-tert-butyl-4-methylphenoxide) (IAB), composed of tris-isobutyl aluminum solution in hexane and 2,6-di-tert-butyl-4-methylphenol.
6. Ground K_2CO_3. Dried overnight at 100 °C in a vacuum oven.
7. Methanol.
8. Fumaryl chloride.
9. Reactor: a three-neck round-bottom glass flask with rubber septum, a nitrogen/vacuum inlet, and thermometer.
10. Schlenk cones (25 mL and 250 mL).
11. Magnetic stirrer.
12. Buchner filtration device.

2.2 Stereolithography

1. N-Vinyl-pyrrolidone (NVP).
2. Lucirin TPO-L (ethyl-2,4,6-trimethylbenzoylphenylphosphinate) (BASF).
3. Orasol Orange G (Ciba Specialty Chemicals).
4. EnvisionTEC Perfactory Mini Multilens stereolithography apparatus, equipped with a digital micro-mirror device which enables projections of 1280x1024 pixels, each measuring 32×32 μm^2 (*see* **Note 1**).

3 Methods

The synthesis of VPCLF should be run using Schlenk lines to ensure a controlled atmosphere that includes any transfer of solvents and reagents to the polymerization reactor. The reactor and other equipment used in the reaction (needles, Schlenk tubes, glass syringes) should be kept overnight in an oven at 135 °C to remove traces of moisture and then quickly assembled and attached to a Schlenk line.

3.1 Synthesis of VPCL

1. Three cycles of vacuum and nitrogen have to be completed before adding the reagents.

2. In a Schlenk 250 mL 19/24 cone (A), dry 5 cm of 4 Å sieves: 3 cycles vac/N$_2$, vacuum and hot gun for 5 min, cooling under vacuum. Then add up to 220 g of ε-caprolactone (20 g lost on sieves).

3. In a 25 mL Schlenk (B), dry 1 cm of sieves (same procedure as above), and add between 10 and 20 mL of HEVE.

4. Prepare 1 dry, cooled 250 mL Schlenk (C) 14/20 cone, with one magnetic stirrer, by assembling hot parts from oven, cooling under vacuum, and run 3 cycles vac/N$_2$. Replace glass stopper with rubber stopper under nitrogen flow.

5. In a three-neck round-bottom flask equipped with a rubber septum, a nitrogen/vacuum inlet, and a thermometer, transfer the required amounts of dry DCM (from half to same as monomer) and ε-CL using a double-tipped needle.

6. In the glove box under argon atmosphere, place the required amount of trimethylaluminum (1 M solution in toluene-catalyst) in a Schlenk tube under magnetic stirring, then slowly add a solution of HEVE in toluene, and keep the mixture under stirring for 10 min.

7. Inject the catalyst precursor through the septum with a nitrogen-flushed syringe to the Schlenk containing the ε-CL (ε-CL/HEVE ratio 12.5:1). Leave the reaction proceed under vigorous stirring overnight, under a static atmosphere of nitrogen at room temperature.

8. Disconnect the flask from the nitrogen inlet and gently pour the content to a cold 1:1 methanol/water bath. Maintain under vigorous stirring for at least 2 h until polymer particles are clearly visible. The bath size needs to be 3–5 times the total volume of the reagents.

9. Filter the methanol suspension gently on a Buchner filtration device and recover the solid VPCL (MW = 1.5 kDa).

10. Dry the solid VPCL overnight in vacuum oven at 30 °C.

3.2 Synthesis of VPCLF

1. Transfer the required amount of previously synthesized VPCL and K$_2$CO$_3$ in a three-neck round-bottom flask, equipped with a magnetic stirrer, a reflux condenser, a rubber septum, and a nitrogen/vacuum inlet, and place in a water bath.

2. Apply three cycles of vacuum/nitrogen to the flask to remove traces of moisture and oxygen, and then transfer dry DCM using a double-tipped needle.

3. After complete dissolution of the polymer, add the desired amount of fumaryl chloride dissolved in dry DCM via injection through the rubber septum to the suspension under vigorous stirring. The reaction is kept overnight at 50 °C under reflux under a static nitrogen atmosphere.

3D design   **2D slicing**   **SLA printing**   **Porous scaffold**

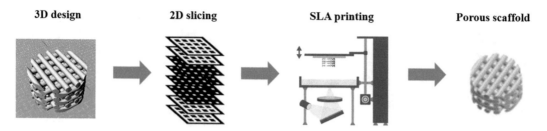

Fig. 1 Schematic overview of the stereolithography process

4. Separate the supernatant from unreacted K_2CO_3, and precipitate the polymer in cold methanol, filter on a Buchner filter, and wash with methanol/water to remove any unreacted chemicals and reaction by-products.

5. Further purify VPCLF by redissolution in DCM and re-precipitation in methanol before drying in a vacuum oven at 30 °C for one night.

3.3 Stereolithography Process

Stereolithography is a layer-by-layer fabrication method, and the schematic overview of the process is presented in Fig. 1. The resins used for stereolithography consist of 50 wt% VPCLF macromer, 50 wt% NVP/Lucrin TPO solution, and 0.2 wt% Orasol Orange G as a dye.

1. Dissolve 5 wt% of Lucirin TPO (relative to VPCLF amount) in NVP by magnetic stirring.

2. Dissolve 1 g of VPCLF in 1 mL of NVP/Lucirin TPO solution by magnetic stirring at room temperature. Add Orasol Orange G dye to the solution (0.15 wt% relative to VPLCF amount), and stir for another 20 min.

3. The thickness of the solidified layer (cure depth, C_d in μm) is controlled by the light irradiation dose E (mJ/cm^2), the penetration depth of the light (D_p), and the critical exposure time (E_c) accordingly with the Beer-Lambert law. A plot of the cure depth versus the irradiation dose (working curve) for VPCLF/NVP resin is reported in Fig. 2 and can be described by the following equation:

$$C_d = D_p \ln \frac{E}{E_c} \qquad (1)$$

4. Gently pour the reactive solution into the SLA resin bath avoiding bubbles formation.

5. To obtain a layer thickness of 25 μm, set the exposure to 20 s with blue light using an intensity of 16 mW/cm^2. Sequentially photocrosslink VPCLF/NVP resins in a layer-by-layer process by exposure to the different light pixel patterns (*see* **Notes 2–4**).

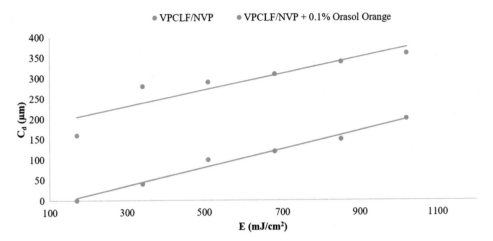

Fig. 2 Relationship between cure depth (C_d) and illumination dose (E) for VPCLF/NVP resin with and without Orasol Orange dye

6. When the built process has finished, extract non-reacted resin by dipping the 3D structures in a 3:1 mixture of isopropanol/ acetone for 30 min.

7. Post-cure porous scaffolds in oven at 90 °C for 24 h under vacuum.

4   Notes

1. Although methods have been optimized for use with this particular SLA equipment, they can easily be adapted to other printers.

2. To design porous 3D structures, TPMS have been adopted with Gyroid (G) and Diamond architectures [23, 24]. K3DSurf v0.6.2 software (http://k3dsurf.sourceforge.net) can be used to generate CAD files that describe the surfaces of Gyroid and Diamond geometries. The trigonometric functions that describe Gyroid and Diamond surface are reported in Eqs. 2 and 3:

$$\cos(x)\sin(y) + \cos(y)\sin(z) + \cos(z)\sin(x) = C \qquad (2)$$

$$\begin{aligned} \sin(x)\sin(y)\sin(z) &+ \sin(x)\cos(y)\cos(z) \\ + \cos(x)\sin(y)\cos(z) &+ \cos(x)\cos(y)\cos(z) \\ &= C \end{aligned} \qquad (3)$$

3. To obtain cylindrical structures with a porosity of ca. 60–70%, an offset value (C) of 0.5 for Gyroid and 0.2 for Diamond can be used with boundary conditions x, $y = [-6\pi; \ 6\pi]$ and $z = [-3\pi; \ 3\pi]$ (Fig. 3).

Unit cell	3D Scaffold	SEM
Gyroid	$x,y = [-6\pi\ ;\ 6\pi]\ \&\ z = [-3\pi\ ;\ 3\pi]$	

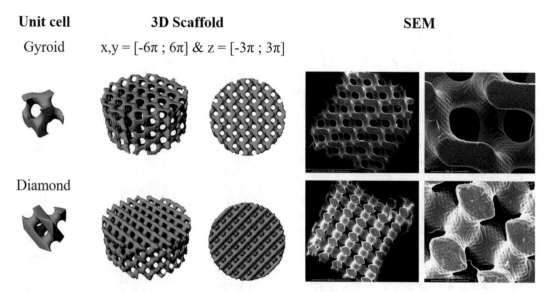

| Diamond | | |

Fig. 3 Visualizations and SEM images of porous structures with Gyroid and Diamond geometry prepared by stereolithography

4. CAD files can be converted to STL files by using Rhinoceros software. The STL files can be sliced into 2D sections, each having 25 μm layer thickness, through the EnvisionTEC Perfactory RP 2.0 software.

Acknowledgments

This work was partially supported by Short Term Mobility Project 2016 (STM 2016) "Photo-crosslinkable polymers for 3D-printing of biomedical devices."

References

1. Yan X, Gu P (1996) A review of rapid prototyping technologies and systems. Comput Aided Des 28(4):307–318

2. Pham DT, Gault RS (1998) A comparison of rapid prototyping technologies. Int J Mach Tools Manuf 38(10–11):1257–1287

3. Kruth JP, Leu MC, Nakagawa T (1998) Progress in additive manufacturing and rapid prototyping. CIRP Ann 47(2):525–540

4. Eloma L, Kokkari A, Närhi T et al (2013) Porous 3D modeled scaffolds of bioactive glass and photocrosslinkable poly (e-caprolactone) by stereolithography. Compos Sci Technol 74:99–106

5. Gloria A, Russo T, De Santis R et al (2009) 3D fiber deposition technique to make multifunctional and tailor-made scaffolds for tissue engineering applications. J Appl Biomater Biomech 7:141–152

6. Rengier F et al (2010) 3D printing based on imaging data: review of medical applications. Int J Comput Assist Radiol Surg 5(4):335–341

7. Petrovic V et al (2011) Additive layered manufacturing: sectors of industrial application shown through case studies. Int J Prod Res 49 (4):1061–1079

8. Potamianos P et al (1998) Rapid prototyping for orthopaedic surgery. Proc Inst Mech Eng H J Eng Med 212(5):383–393

9. Puppi D, Chiellini F, Piras AM, Chiellini E (2010) Polymeric materials for bone and cartilage repair. Prog Polym Sci 35(4):403–440

10. Mota C, Puppi D, Chiellini F, Chiellini E (2015) Additive manufacturing techniques for the production of tissue engineering constructs. J Tissue Eng Regen Med 9:174–190

11. Melchels FPW, Bertoldi K, Gabbrielli R et al (2010) Mathematically defined tissue engineering scaffold architectures prepared by stereolithography. Biomaterials 31(27):6909–6916

12. Ronca A, Ambrosio L, Grijpma DW (2012) Design of porous three dimensional PDLLA/ nano-hap composite scaffolds using stereolithography. J Appl Biomater Funct 10 (3):249–258

13. Melchels FPW, Feijen J, Grijpma DW (2010) A review on stereolithography and its applications in biomedical engineering. Biomaterials 31:6121–6130

14. Raman R, Rashid B (2015) Stereolithographic 3D bioprinting for biomedical applications. In: Essentials of 3D biofabrication and translation. Academic Press, Cambridge, pp 89–121

15. Gauvin R, Chen YC, Lee JW et al (2012) Microfabrication of complex porous tissue engineering scaffolds using 3D projection stereolithography. Biomaterials 33 (15):3824–3834

16. Ronca A, Ambrosio L Grijpma DW (2013) Preparation of designed poly(d,l-lactide)/ nanosized hydroxyapatite composite structures by stereolithography. Acta Biomater 9:5989–5996

17. Melchels FPW, Feijen J, Grijpma DW (2009) A poly(d,l-lactide) resin for the preparation of tissue engineering scaffolds by stereolithography. Biomaterials 30:3801–3809

18. Grijpma DW, Qingpu H, Feijen J (2005) Preparation of biodegradable networks by photo-crosslinking lactide, ε-caprolactone and trimethylene carbonate-based oligomers functionalized with fumaric acid monoethyl ester. Biomaterials 26(16):2795–2802

19. De Santis R, Gloria A, Prisco D et al (2010) Fast curing of restorative materials through the soft light energy release. Dent Mater 26:891–900

20. Yoshii E (1997) Cytotoxic effects of acrylates and methacrylates: relationships of monomer structures and cytotoxicity. J Biomed Mater Res 37:517–524

21. Liow SS, Widjaja LK, Lipik VT et al (2009) Synthesis, characterization and photopolymerization of vinyl functionalized poly (-ε-caprolactone). Express Polym Lett 3 (3):159–167

22. Ronca A et al (2018) Synthesis and characterization of divinyl-fumarate poly-ε-caprolactone for scaffolds with controlled architectures. J Tissue Eng Regen Med 12(1):523–531

23. Gandy PJ, Klinowski J (2000) Exact computation of the triply periodic G (Gyroid') minimal surface. Chem Phys Lett 321(5–6):363–371

24. Gandy PJF et al (1999) Exact computation of the triply periodic D (diamond') minimal surface. Chem Phys Lett 314(5–6):543–551

Chapter 6

Nanocomposite Clay-Based Bioinks for Skeletal Tissue Engineering

Gianluca Cidonio, Michael Glinka, Yang-Hee Kim, Jonathan I. Dawson, and Richard O. C. Oreffo

Abstract

Biofabrication is revolutionizing substitute tissue manufacturing. Skeletal stem cells (SSCs) can be blended with hydrogel biomaterials and printed to form three-dimensional structures that can closely mimic tissues of interest. Our bioink formulation takes into account the potential for cell printing including a bioink nanocomposite that contains low fraction polymeric content to facilitate cell encapsulation and survival, while preserving hydrogel integrity and mechanical properties following extrusion. Clay inclusion to the nanocomposite strengthens the alginate-methylcellulose network providing a biopaste with unique shear-thinning properties that can be easily prepared under sterile conditions. SSCs can be mixed with the clay-based paste, and the resulting bioink can be printed in 3D structures ready for implantation. In this chapter, we provide the methodology for preparation, encapsulation, and printing of SSCs in a unique clay-based bioink.

Key words Biofabrication, Bioink, Clay, Laponite, Scaffolds, Bone repair, Skeletal stem cell

1 Introduction

Clay has been commonly used in the pharmaceutical industry (predominantly as excipients with roles as lubricants, diluents, flavor correctors, emulsifiers, rheological agents, and drug delivery modifiers in areas from gastroenterology, antacids, and antidiarrheics to aesthetic medicine and cosmetics) [1, 2] and in the last few years has generated significant interest as a biomaterial for regenerative medicine [3].

Clay minerals, also called sheet silicates or phyllosilicates, are a family of inorganic layered nanomaterials of which smectites, specifically Laponite, are most relevant for biomaterial design and bioink development. Laponite is composed of disk-like nanoparticles 25 nm in diameter and 1 nm in thickness with negatively charged faces and a weak positively charged rim surface

Alberto Rainer and Lorenzo Moroni (eds.), *Computer-Aided Tissue Engineering: Methods and Protocols*, Methods in Molecular Biology, vol. 2147, https://doi.org/10.1007/978-1-0716-0611-7_6, © Springer Science+Business Media, LLC, part of Springer Nature 2021

[4]. Dispersion of Laponite in an aqueous environment results in a colloidal suspension of clay nanoparticles, which can associate to form gels via both repulsive and attractive interactions depending on solid concentration and the ionic concentration of the solution [5, 6].

Laponite has several functional properties, including (1) shear thinning and thixotropic behavior [7], (2) retention or localization of drugs due to strong protein-clay interactions [8], and (3) cell stimulation [9], as well as its widespread use in polymer clay nanocomposites. Due to the physical nature of Laponite interactions, nano-colloids and composites often display shear thinning or self-healing properties which can be particularly useful for cell printing processes [10].

Bioinks (biomaterials blended with living cells) have recently come to the fore in regenerative medicine, given their versatility and functional potential to deliver living cells in 3D, within discrete tailored architecture and/or highly ordered cell-laden structures. Biofabrication can generate large-scale structures with a plethora of biomaterials. Nevertheless, bioinks that can elicit a degree of functional stimulation toward the encapsulated cells remain to be developed. Clay-based hydrogels can be tuned to retain or localize drugs after brief exposure, stimulating angiogenesis [8] or bone formation [11] illustrating the clinical potential and opportunity provided by Laponite and related materials to provide a one-step procedure to encapsulate, retain, and deliver therapeutic agents [8]. Moreover, recent advances [9] have investigated the direct effect that Laponite nanodispersion can have on human mesenchymal stem cells (hMSCs), identifying a significant influence on over 4000 genes and major cellular pathways including mitogen-activated protein kinase (MAPK).

Laponite functionality has been used in combination with several polymers to control mechanical properties of the gel [12], drug delivery [13], cell attachment [14], and detachment [15]. Importantly, clay integration within polymer matrices provides exceptional shear-thinning modification essential for printing applications. We have developed a functional clay-based bioink by blending Laponite with alginate and methylcellulose in 3% w/v concentration (3-3-3 for short) encapsulating stem cells that can be printed with high shape fidelity and preserving cell viability for up to 21 days in culture (Figs. 1–3).

2   Materials

2.1 Clay-Based Biopaste

1. Sterile deionized water (DW, conductivity 18.2 MΩ cm at 20 °C) (*see* **Note 1**).

Fig. 1 Laponite suspension. Immediately after Laponite addition (0 s) to deionized water, nanosilicate particles typically result in an opaque solution. After 20 s, Laponite suspension will start to clear, demonstrating Laponite nanoparticles dispersal. By 50 s, Laponite dispersal achieved, evidenced by a water-clear solution. Laponite should be left stirring continuously for at least 3 h (*see* **Note 2**) to allow complete dispersion of the nanoparticles in suspension

Fig. 2 Addition of methylcellulose to Laponite-alginate suspension. Immediately after removal of magnetic stirrer bar, methylcellulose powder can be directly mixed using a sterile spatula (**a**) addition of methylcellulose powder deposited on top of the suspension (**b**). Resultant mixture (**c**) after at least 1 min of stirring with a spatula, the mixture can be stored at 4 °C overnight to allow methylcellulose homogenous inclusion

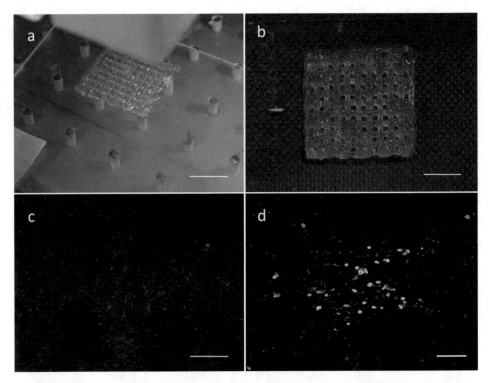

Fig. 3 3D printing of cell-laden 3-3-3 bioink. Printing of 3-3-3 (**a**) with an in-house-built 3D bioprinter using a conical 410 μm nozzle. Crosslinking after printing can be performed with immersion in 100 mM CaCl$_2$ solution. 12 × 12 mm construct (**b**) can then be used in biological studies in vitro. Cell-laden scaffold (**c**) viability can be evaluated at specific time points. To aid cell visualization (using a fluorescence microscope), DiD (red) preprinting labelling of the cells can be performed. Close-up of cell-laden strand (**d**) with cells labelled with DiD (red) and viable cells labelled with calcein AM (green). Scale bars: (**a**, **b**) 5 mm, (**c**) 1 mm, (**d**) 100 μm

2. Laboratory glass bottles with caps (25 mL, Duran) to store biomaterial powder.

3. Glass beaker (20 or 40 mL volume capacity, depending on the amount of biopaste needed).

4. Magnetic stirrer bar.

5. Laponite XLG (BYK Additives & Instruments).

6. Alginate (alginic acid sodium salt from brown algae, mannuronic/glucuronic (M/G) acid ratio 1:2).

7. Methylcellulose (MW ≈ 88 kDa, 4000 cP).

8. Sterile stainless steel spatulas.

9. Magnetic stirrer (reaching 28 RCF to 300 RPM).

2.2 Cell Culture

1. Skeletal stem cells (SSCs; isolated from human sources with full national ethical approval and after patient consent acquired).

2. Cell culture flasks (175 cm^2 cell culture flask, angled neck, non-pyrogenic polystyrene).

3. Phosphate buffer saline (Dulbecco's phosphate-buffered saline (PBS) without calcium and magnesium).

4. Complete medium: α-MEM (alpha-minimum essential medium with L-glutamine) containing 10% fetal bovine serum (FBS) and 1% penicillin-streptomycin mixture (10,000 U/mL penicillin/mL, 10,000 U/mL streptomycin).

5. Collagenase working solution: prepare a 200 µg/mL solution of collagenase (collagenase type I from *Clostridium histolyticum*, sterile-filtered) in serum-free α-MEM.

6. Trypsin/EDTA (trypsin/ethylenediaminetetraacetic acid, for cell culture).

7. Vybrant DiD (Thermo Fisher Scientific) cell-labelling solution.

2.3 Bioprinting

1. Luer-lock syringe, 10–20 mL volume.

2. Sterile spatula for paste loading.

3. Luer-lock nozzle tips (22 gauge tapered dispensing tip (ID 0.41 mm) (sterilized in ethanol or by gamma-irradiation).

4. Extrusion-based bioprinter, able to generate 80–95 kPa pressure on biopaste (biopaste has been tested in pneumatic [16] and piston-driven extrusion printer).

5. Six-well tissue culture plate or, alternatively, 35 mm Petri dishes.

2.4 Post-printing Processing

1. Calcium chloride solution: 100 mM calcium chloride ($CaCl_2$) in DW.

2. Sterile spatulas.

3. Sterile 12-well plates containing α-MEM to transfer scaffolds after crosslinking.

4. Hank's Balanced Salt Solution (HBSS, calcium, magnesium).

3 Methods

Perform the following procedures at room temperature in a cell culture hood with laminar flow when synthesizing biopaste for cell printing.

3.1 Biopaste Synthesis

1. Separately autoclave proper amounts of biomaterial powders (Laponite, alginate, and methylcellulose) in glass bottles with lid.

2. Add 20 mL DW to a sterile beaker with magnetic bar.

3. Keeping the solution under magnetic stirring at 14 RCF, slowly add 600 mg of sterile nanosilicate Laponite powder. Perform a slow and gentle addition of the powder, pausing

when the stirring suspension starts becoming opaque and resuming when it has cleared out.

4. Leave Laponite to fully disperse in suspension by continuous stirring for at least 3 h (Fig. 1; *see* **Note 2**). Alginate powder (600 mg) can then be added to the stirring suspension by slow addition. Alginate dispersion will be completed by stirring the suspension for 2 h (*see* **Note 3**).

5. Lastly, methylcellulose will be added. Carefully remove the stirring bar with a sterile spatula, avoiding collecting the suspension and removing excessive Laponite-alginate hydrogel from the stirrer bar. Insert 600 mg of methylcellulose in the static Laponite-alginate suspension. Use a sterile spatula (Fig. 2a) to stir and disperse the methylcellulose powder close to the top of the suspension (Fig. 2b). Once the methylcellulose powder is dispersed (Fig. 2c), store the 3-3-3 suspension at 4 °C overnight (*see* **Note 4**).

3.2 Preparation of Stem Cells for Printing

1. Skeletal stem cells can be expanded in T175 cm^2 flasks to reach desired number for studies to be undertaken (*see* **Note 5**).

2. Allow cell culture media, PBS, collagenase solution, and trypsin/EDTA to warm up at least 30 min.

3. Remove cell culture media from flasks, wash with PBS, and insert 10 mL of collagenase working solution in one T175 cm^2 flask. Incubate for 45 min at 37 °C, 5% CO_2 (*see* **Note 6**).

4. Remove collagenase media, add trypsin/EDTA, and incubate for 5 min at 37 °C, 5% CO_2. Tap the flasks to allow full stem cell detachment (*see* **Note 7**). Rinse flask with complete α-MEM, to deactivate trypsin, and collect suspension in a falcon tube. Centrifuge at 228 RCF for 4 min. Discard supernatant and resuspend pellet to count stem cells.

5. Cells can be fluorescently labelled using a live-compatible stain, such as Vybrant DiD (Thermo Fisher Scientific). Prepare a staining solution at a concentration of 5 μL/mL in FBS-free medium, and resuspend cells at 1×10^6 cells/mL. Incubate (37 °C, 5% CO_2) for 20 min, and then wash three times with fresh complete cell culture media.

6. Allow $1–5 \times 10^6$ cells per gram of clay-based biopaste (*see* **Note 8**). Collect the required amount of cell suspension, centrifuge, and pellet cells.

3.3 Cell-Laden Clay-Based Bioinks Printing

1. In a sterile cell culture hood, load 3-3-3 in a pre-weighed falcon tube. Weigh the paste to obtain 5 g of material. Spin down the 3-3-3 at 755 RCF for 1 min (*see* **Note 9**).

2. Resuspend cell pellet in 100 μL complete α-MEM media, collect cells, and pipette the suspension on top of the 3-3-3.

Use a sterile spatula to gently mix thoroughly for at least 1 min (*see* **Note 10**).

3. Load the cell-laden clay-based bioink in a Luer-lock syringe. 10 or 20 mL syringes are to be preferred, as they have a wider aperture to allow easier biopaste loading (*see* **Note 11**).

4. Syringe plunger can be inserted, pushing 3-3-3 toward the tip. Lock in a sterile Luer-lock conical nozzle of correct dimension for cell printing (*see* **Note 12**).

5. Wipe 3D printer depositing surface, syringe holder, and piston with 70% ethanol for sterile cell printing.

6. Lattice 0/90° can be readily printed (Fig. 3a) with stem cell-laden 3-3-3 using a pressure between 80 and 95 kPa, depending on printer setup, pressure-driven adaptors, pressure system, and printer mechanics (*see* **Note 13**).

7. Several structures can be produced by 3-3-3 printing as detailed in our recent publication [16]. A 12×12 mm construct (Fig. 3b, c) can be readily printed with a single-arm piston-driven extrusion-based bioprinter (*see* **Note 14**).

8. Post-printing, crosslink scaffolds with calcium chloride solution for 10 min. Remove calcium chloride solution, and cultivate cell-laden scaffolds in complete medium for long-term culture at 37 °C, 5% CO_2, changing medium every 2 days (*see* **Note 15**).

9. If cell viability investigation is needed, this can be performed with calcein-AM staining. To stain constructs at time points, use 0.6 μL calcein AM per mL of FBS-free medium. Incubate (37 °C, 5% CO_2) for 1 h. Remove medium, and wash with HBSS at least twice. Image with fluorescence or confocal microscope (Fig. 3d).

---

4 Notes

1. It is crucial for clay-based composite hydrogels to be prepared from DW with specific resistivity (18.2 MΩ cm) showing signs of coagulation or precipitation at higher and lower resistivity value, respectively.

2. It is crucial that Laponite suspension is left for 3 h to stir continuously at room temperature at 14 RCF. This allows Laponite nanodiscs to disperse homogenously resulting in a clear solution. If Laponite is not allowed to disperse, it will aggregate and eventually fail to intercalate with alginate and methylcellulose polymeric chain.

3. Powder may aggregate at the air-liquid interface. By momentarily raising the stirring speed to 28 RCF, the alginate

aggregation will be included within the suspension, although constant stirring at 14 RCF will be needed immediately after. If alginate inclusion results in aggregations after 1 h of stirring, a sterile-autoclaved spatula can be used to break down alginate aggregations. Mixing will then need to be resumed for further 2 h.

4. After methylcellulose addition and stirring with a sterile spatula, some methylcellulose aggregates will still be visible. Overnight storage at 4 °C will facilitate full methylcellulose inclusion, resulting in a homogeneous amber-colored biopaste ready to be printed.

5. SSCs expansion and density number will depend on donor, passage, and culture time. We strongly recommend using SSCs at low passage numbers (Passage 1 or Passage 2) to preserve SSCs stemness and avoid uncontrolled differentiation prior to the experiment.

6. Incubation time will vary depending on specific SSCs and donor used. Forty-five minutes is the average time of SSCs incubation within the collagenase solution.

7. Confirmation of full detachment should be performed with a light microscope.

8. Cell density for inclusion in bioinks is still debated. We have previously used 1×10^6 cells/g, but this value can be increased or lowered depending on the experimental plan (viability, proliferation, in vivo).

9. This procedure is carried out to obtain a paste that can be easily loaded and cells can be quickly mixed. Fast spinning is recommended.

10. Mixing must be carried out gently. Using a sterile stainless steel spatula, stem cells can be easily mixed within paste by slow clockwise bottom-up mixing. This procedure is critical for post-printing viability and must be performed carefully.

11. A sterile spatula can be used to scoop 3-3-3 from the beaker and slowly adding it to the syringe barrel. For a piston-driven printing process, using a Luer-lock disposable syringe, it is recommended that biopaste should not be loaded more than ¾ of the syringe total volume.

12. Conical-shaped disposable sterile nozzles are preferable when printing living cells [17]. Flow-dynamic force field representing shear stresses found during extrusion indicates higher forces applied to nozzle boundaries when a cylindrical extruder is used. Conical nozzles show a lower degree of shear applied along the extrusion direction, resulting in a clear advantage in preserving cell viability.

13. Pressure needed to extrude cell-laden 3-3-3 paste may vary according to the bioprinter used. A test with acellular printing of 3-3-3 is recommended before using the cell-laden paste.

14. Cell viability, proliferation, and functionality studies typically need a large amount of constructs. Stem cell-laden 3-3-3 can be printed as a high-throughput process.

15. After crosslinking solution removal, a single wash with complete cell culture media can be performed to remove excess $CaCl_2$ solution. Use enough cell culture media to completely cover the entirely cell-laden scaffold.

Acknowledgments

The authors would like to thank Prof. Michael Gelinsky (TU Dresden) for useful discussions and fruitful collaborations over the last 3 years, Prof. Shoufeng Yang (KU Leuven) for discussions and access to the extrusion bioprinter, and Dr. Stuart Lanham for useful discussions on methods. This work was supported by grants from the Biotechnology and Biological Sciences Research Council UK (BB/ L00609X and BB/LO21072/1) and University of Southampton IfLS, FortisNet and Postgraduate awards to ROCO.

References

1. Carretero MI, Pozo M (2009) Clay and non-clay minerals in the pharmaceutical industry. Part I. Excipients and medical applications. Appl Clay Sci 46:73–80

2. Carretero MI, Pozo M (2010) Clay and non-clay minerals in the pharmaceutical and cosmetic industries Part II. Active ingredients. Appl Clay Sci 47:171–181

3. Dawson JI, Oreffo ROC (2013) Clay: New opportunities for tissue regeneration and biomaterial design. Adv Mater 25:4069–4086

4. Ruzicka B, Zaccarelli E (2011) A fresh look at the Laponite phase diagram. Soft Matter 7:1268–1286

5. Kroon M, Vos WL, Wegdam GH (1998) Structure and formation of a gel of colloidal disks. Int J Thermophys 19:887–894

6. Abou B, Bonn D, Meunier J (2001) Aging dynamics in a colloidal glass. Phys Rev E Stat Phys Plasmas Fluids Relat Interdiscip Topics 64:6

7. Pignon F, Magnin A, Piau JM (1998) Thixotropic behavior of clay dispersions: combinations of scattering and rheometric techniques. J Rheol 42:1349–1373

8. Dawson JI, Kanczler JM, Yang XB et al (2011) Clay gels for the delivery of regenerative microenvironments. Adv Mater 23:3304–3308

9. Carrow JK, Cross LM, Reese RW et al (2018) Widespread changes in transcriptome profile of human mesenchymal stem cells induced by two-dimensional nanosilicates. Proc Natl Acad Sci U S A 115(17):E3905–E3913

10. Hölzl K, Lin S, Tytgat L et al (2016) Bioink properties before, during and after 3D bioprinting. Biofabrication 8:032002

11. Gibbs DMR, Black CRM, Hulsart-Billstrom G et al (2016) Bone induction at physiological doses of BMP through localization by clay nanoparticle gels. Biomaterials 99:16–23

12. Liu X, Bhatia SR (2015) Laponite® and Laponite®-PEO hydrogels with enhanced elasticity in phosphate-buffered saline. Polym Adv Technol 26:874–879

13. Viseras C, Aguzzi C, Cerezo P et al (2008) Biopolymer–clay nanocomposites for controlled drug delivery. Mater Sci Technol 24:1020–1026

14. Gaharwar AK, Schexnailder PJ, Kline BP et al (2011) Assessment of using Laponite cross-

linked poly(ethylene oxide) for controlled cell adhesion and mineralization. Acta Biomater 7:568–577

15. Haraguchi K, Takehisa T, Ebato M (2006) Control of cell cultivation and cell sheet detachment on the surface of polymer/clay nanocomposite hydrogels. Biomacromolecules 7:3267–3275

16. Ahlfeld T, Cidonio G, Kilian D et al (2017) Development of a clay based bioink for 3D cell printing for skeletal application. Biofabrication 9:034103

17. Billiet T, Gevaert E, De Schryver T et al (2014) The 3D printing of gelatin methacrylamide cell-laden tissue-engineered constructs with high cell viability. Biomaterials 35:49–62

Part III

Technological Platforms and Manufacturing Processes

Chapter 7

Additive Manufacturing Using Melt Extruded Thermoplastics for Tissue Engineering

Andrea Roberto Calore, Ravi Sinha, Jules Harings, Katrien V. Bernaerts, Carlos Mota, and Lorenzo Moroni

Abstract

Melt extrusion of thermoplastic materials is an important technique for fabricating tissue engineering scaffolds by additive manufacturing methods. Scaffold manufacturing is commonly achieved by one of the following extrusion-based techniques: fused deposition modelling (FDM), 3D-fiber deposition (3DF), and bioextrusion. FDM needs the input material to be strictly in the form of a filament, whereas 3DF and bioextrusion can be used to process input material in several forms, such as pellets or powder. This chapter outlines a common workflow for all these methods, going from the material to a scaffold, while highlighting the special requirements of particular methods. A few ways of characterizing the scaffolds are also briefly described.

Key words Tissue engineering, Scaffolds, 3D printing, Fused deposition modelling, Bioextrusion, 3D-fiber deposition, Biofabrication

1 Introduction

Additive manufacturing (AM) is a group of well-established techniques in industry to rapidly manufacture objects in a layer-by-layer manner where each individual layer is also created progressively, adding material rather than removing it. Since the last three decades, it has been applied in tissue engineering (TE) to manufacture scaffolds with very complex shapes, thanks to its high degree of control on architectural parameters such as pore size, pore shape, and porosity [1, 2].

Depending on the material to be processed and on the processing method itself (which are highly interrelated), several AM techniques have been developed over the last decades. A possible

Electronic supplementary material: The online version of this chapter (https://doi.org/10.1007/978-1-0716-0611-7_7) contains supplementary material, which is available to authorized users.

Alberto Rainer and Lorenzo Moroni (eds.), *Computer-Aided Tissue Engineering: Methods and Protocols*,
Methods in Molecular Biology, vol. 2147, https://doi.org/10.1007/978-1-0716-0611-7_7,
© Springer Science+Business Media, LLC, part of Springer Nature 2021

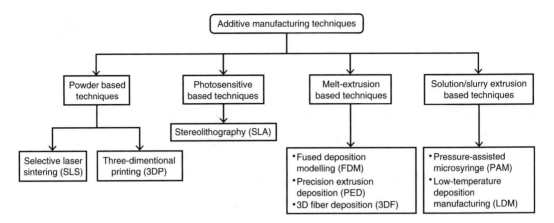

Fig. 1 Proposed classification of the AM techniques commonly employed in TE (adapted from [2])

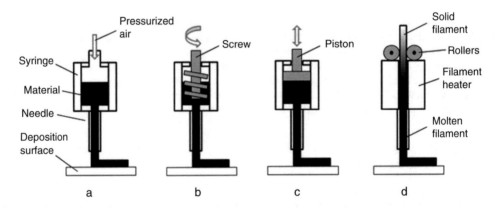

Fig. 2 Schematic of fluid dispensing approaches, (**a**) gas pressure, (**b**) rotary screw, (**c**) positive displacement, and (**d**) roller-assisted (readapted from [3]). In the first three techniques, the material is melted and then extruded by means of pressurized gas (**a**), a screw (**b**), or a piston (**c**). In (**d**), a solid filament, as it is pushed through a heater (also known as liquefier) by rollers, is melted and can be deposited on the deposition surface

classification is shown in Fig. 1. Selective laser sintering (SLS) uses a laser to selectively sinter powdered material on a heated bed rendering the processing of metals, ceramics, and polymers. Stereolithography (SLA) works by selectively photopolymerizing layers of a photosensitive polymer by means of an ultraviolet (UV) light or laser; both metals and polymers are used. Three-dimensional printing (3DP) is based on the controlled deposition of a binder material laid on a powder layer using an inkjet head, and it is used with polymers. In extrusion-based techniques, the material is made to flow through a nozzle upon the application of pressure via pressurized gas/piston (as shown in Fig. 2a, c), via a screw (Fig. 2b), or via rollers pushing a filament (that acts as a piston, Fig. 2d). The input material is either in molten state (particularly for polymers) or in a slurry form (polymers in solution or hydrogels). The group of techniques based on molten materials is usually referred to as fused deposition modelling (FDM) even though this is the name of the filament-based approach [2, 4]. As different technologies, still based on melt extrusion, were developed over time, a new nomenclature was recently suggested. In particular, the pressurized gas/piston-based approach is now being referred to as 3D-fiber deposition, while bioextrusion is used for screw-driven processes [5].

Although each approach has its own advantages, the type of material that can be used with a particular technique is limited by the manufacturing process. For example, FDM and SLS require polymers that become moldable above a specific temperature and solidify upon cooling (called thermoplastics), SLA requires the use of a photopolymer resin, and 3DP involves the use of solvents and binders [6, 7].

Tissue engineers have focused their attention on AM techniques attracted by the possibility of controlling the geometry of the scaffold, an essential parameter to be considered when mimicking a tissue's morphology and mechanical properties. Principles and materials typically used in industry have been adapted to the stricter requirements of tissue engineering, mainly due to biocompatibility, but also to the different product morphologies involved, less dense and with a high level of porosity.

Although the first machines developed starting from the 1950s were based on SLA, FDM techniques are probably the most successful ones due to their simplicity and low demands on hardware, which makes low-cost printers possible [8]. This applies to the field of tissue engineering in particular, where the high costs of medical grade materials and of the process toward clinical application are pushing research groups to use extrusion-based machines. With the only theoretical requirement for the material to be extrudable, the devices are usually simpler and the operational risks lower compared to SLA, SLS, and 3D printing. This is due to the fact that, generally, no toxic precursors or binders are needed during the shaping process. Additionally, when working with polymers as for hard tissue engineering (e.g., bone and cartilage), melt-based techniques are preferred to solution-based extrusion as they do not involve any solvent that might be toxic for cells. As the polymer cools down and solidifies after extrusion, its stiffness and mechanical strength are high enough so that no curing/gelation step is needed (usually needed with hydrogels). The polymer also retains the given shape as no solvent evaporates.

In tissue engineering, extrusion techniques have been successfully used to produce scaffolds made of hydrogels as well as thermoplastic polymers. The use of hydrogels allows the user to avoid solvents or high temperatures for the extrusion, thanks to their viscoelastic behavior. Therefore, tissue engineers have started embedding cells in the material to be extruded to avoid the further step of cell seeding. Additionally, a better controlled spatial distribution of cells can be achieved. Soft tissues such as skin or muscle have been engineered by using hydrogels [9].

On the other hand, the regeneration of stiffer tissues, such as bone and cartilage, has been investigated through the extrusion of thermoplastic polymers. This class of materials naturally offers higher mechanical properties compared to hydrogels, even though the biological cues and the affinity for cells are more limited. To extrude thermoplastics, either solvents or high temperatures are needed, and thus cells cannot be included directly into the printing process. Nevertheless, the mechanical properties have been shown to be suitable for the regeneration of stiff tissues such as cartilage and bone. In particular, it was reported that cartilage mechanical properties could be successfully mimicked by adjusting the fiber deposition pattern of a molten polymer with a pressure-based printer [10, 11]. The same machine was used to produce also composite scaffolds made of a thermoplastic polymer and hydroxyapatite for bone tissue regeneration [12]. Fibroblasts were successfully cultured on scaffolds made of a thermoplastic polymer/composite material with a filament-based machine [13]. Recently, a scaffold platform that can actively boost vascularization and may be applied for extrahepatic islet transplantation was manufactured with a thermoplastic polymer deposited with a screw-based extrusion machine [14]. The same approach was used to fabricate scaffolds with an in-built radial interconnected porosity gradient, and cell differentiation assays confirmed the differentiation of hMSCs toward the osteogenic lineage [15].

Considering the huge potential of extrusion-based printing shown in literature, here we demonstrate that scaffolds with the desired geometry, filament diameter, and filament orientation can be successfully produced by means of the three different approaches. We show how to determine the right printing parameters for each technique according to the material properties, and we discuss about the advantages and drawbacks of each approach to be considered before printing.

2 Materials

1. PolyActive 300PEOT55PBT45 (molecular weight of the initial PEG used for copolymerization, 300 g/mol; block weight ratio PEOT/PBT, 55/45). The material is kept stored in a dark and dry place in vacuum-sealed bags prior to use to avoid contamination and moisture absorption.

2. Differential scanning calorimetry (DSC) pans.

3. Microbalance.

4. DSC apparatus.

5. Rheometer.

6. Computer equipped with CAD software.

7. Mini extruder equipped with 1.5 mm die.

8. Pushing rod.

9. Aluminum foil.

10. Vacuum bags.

11. Vacuum sealer.

12. Fused deposition modelling apparatus and relative controlling and slicing softwares.

13. 3D-fiber deposition apparatus and relative controlling and slicing softwares.

14. Bioextrusion apparatus and relative controlling and slicing softwares.

15. Double-sided tape.

16. 70% ethanol

17. SEM sample holders and carbon conductive tape.

18. Liquid nitrogen.

19. Razor blade.

20. SEM apparatus.

21. Sputter coater apparatus (including argon and gold reservoirs).

3   Methods

The methods described below apply to all three formats of thermoplastic printing, namely, fused deposition modelling (FDM), 3D-fiber deposition (3DF), and bioextrusion, unless differently specified. First, the thermal transitions of the material were analyzed via DSC to determine the temperature region where the polymer is in molten state and can therefore be processed. Then rheometry was used to simulate the flow behavior during printing and to understand the polymer response to processing conditions.

3.1   Material Characterization

Insights about the thermal and rheological properties of a thermoplastic material are very useful for determining its processing conditions. Two material characterization tests that provide such useful insights are differential scanning calorimetry (DSC) and rheometry.

3.1.1   DSC

The DSC measurements described here were run under nitrogen atmosphere and by using a hermetically sealed empty aluminum pan as a reference. The resulting plot for 300PEOT55PBT45 is shown in Fig. 11. To increase the accuracy of the measurements, *see* **Notes 1–3**:

1. Find the combined weight of one aluminum pan and one hermetic lid.

2. Cut one pellet of material into a slice of around 5 mg, weigh it, and place it inside the pan with the cross section lying on the bottom.

3. Seal the pan with the hermetic lid and load the pan in the DSC apparatus.

4. Program the software to measure the sample according to the following procedure:

   (a) Equilibrate the sample temperature at 25 °C.

   (b) Heat up to 195 °C at a rate of 10 °C/min. The final temperature of the heating cycle should be chosen considering the issues explained in **Note 4**.

   (c) Keep the material isothermally for 3 min and then cool to 25 °C at the same rate. A second heating cycle might be needed as described in **Note 5**.

3.1.2 Rheometry

Rheometry measurements were run under nitrogen atmosphere and using a 25 mm plate-plate geometry. The results are shown in Figs. 12 and 13:

1. Set the temperature of the oven at 200 °C and allow it to stabilize.

2. Calibrate the gap and load some material on the bottom plate. This step should be carried out taking into account the precautions described in **Note 6**.

3. Allow the temperature to stabilize again and then close the plates to the measuring gap. Trim the extra material and lower the gap by an additional 15 μm to get a proper meniscus. Consider the recommendations in **Notes 7** and **8** for proper gap adjusting.

4. Start the measurement according to the following protocol:

   (a) Time sweeps at 1% strain, 1 rad/s, for 300 s.

   (b) Frequency sweep at 1% strain, from 628 rad/s to 0.1 rad/s. When choosing the frequency range of interest, consider the risk of outflow as described in **Note 9**.

   (c) Temperature ramp from the current temperature to 20 °C lower, at 1% strain, 1 rad/s and 5 °C/min.

   (d) Repeat **steps 2** and **3** until complex modulus G* exceeds 10^7 Pa.

   (e) Temperature ramp from current temperature to 200 °C, at 1% strain, 1 rad/s and 5 °C/min.

   (f) Frequency sweep at 1% strain, from 628 rad/s to 0.1 rad/s.

3.2 Geometry Preparation

The geometry consisted in a prism of squared base. The CAD model was designed with a commercially available CAD software:

1. Sketch a square of 2×2 cm.

2. Extrude it along the vertical axis for 0.4 cm.

3. Save the file in .STL format.

3.3 Start Fabrication

The techniques of choice used to fabricate scaffolds were fused deposition modelling (FDM), 3D-fiber deposition (3DF), and bioextrusion. As a representative case, the machines used were, respectively, Hyrel 3D 30M, EnvisionTEC V1.0 Bioplotter, and SysEng BioScaffolder. All three machines require some preliminary steps before actively manufacturing. These include assembling the extrusion head (*see* **Note 10**), calibrating the working distance (*see* **Note 11**), preparing the code, and loading the material. The devices can be seen in Fig. 3, while Fig. 4 shows a picture of the extrusion heads and the nozzles used. For a quality assessment, check the scaffolds under a light microscope, and adjust the printing parameters according to the specific technique used (temperature, pressure or screw speed or rollers speed, deposition speed) to get the desired quality, and manufacture again (*see* **Note 12** for further information about the choice of the right parameters). Table 1 lists a set of optimized parameters for the materials described here. A summary of all the main features of the three different techniques can be found in Table 3.

3.3.1 Fused Deposition Modelling

Printing with Hyrel 3D 30M printer (Fig. 3c) is described here as an example of FDM. To manufacture the thermoplastic material necessary for this printer, a step on conversion of the pellets into filament is performed using a twin-screw extruder:

1. Mount the 1.5 mm die.

2. Set the temperature of the machine to 140 °C and wait for it to stabilize. *See* **Notes 13–15** for recommendations about the choice of the right temperature for deposition.

3. Start the motor by setting the rpm to 100.

4. Pour progressively up to 7 g of material in the hopper and push it with the specific tool.

5. Mix for 5 min, open the bypass valve, and collect the filament extruded through the die on aluminum foil, avoiding the different sections to fuse together.

6. Cut any nonuniform sections of the filament.

7. Store the filament in a vacuum-sealed bag in a dry and dark place.

To perform the scaffold fabrication with the Hyrel AM System, install first the two-roller extruder head (Fig. 4c left) and set the

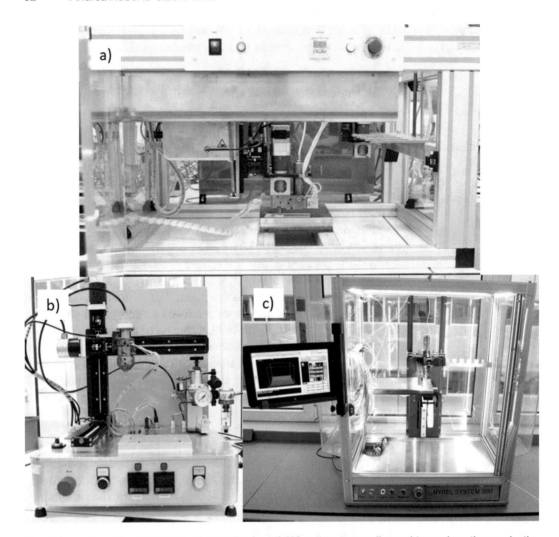

Fig. 3 Examples of the three types of extrusion-based AM systems normally used to produce thermoplastic-based scaffolds: (**a**) SysEng BioScaffolder with a screw extrusion-based head assisted by pressurized reservoir (bioextrusion example); (**b**) EnvisionTEC V1.0 Bioplotter, which uses gas pressure-based extrusion (3DF example); and (**c**) Hyrel 3D printer with a filament-based head (FDM example)

printing height. Lastly, load the filament (brief visual instructions can be found in Supplementary Video 1):

1. Mount the 500 μm nozzle (Fig. 4c right) to the extrusion head.

2. Slide the extruder gibs (Fig. 5a) into the head gib slot (Fig. 5b), and make sure the extruder connector is seated properly (Fig. 5c).

3. Tighten the thumbscrew to lock the extruder (Fig. 5d).

4. Stick a stripe of double-sided tape on the building platform (Fig. 5e).

5. Start the controlling software.

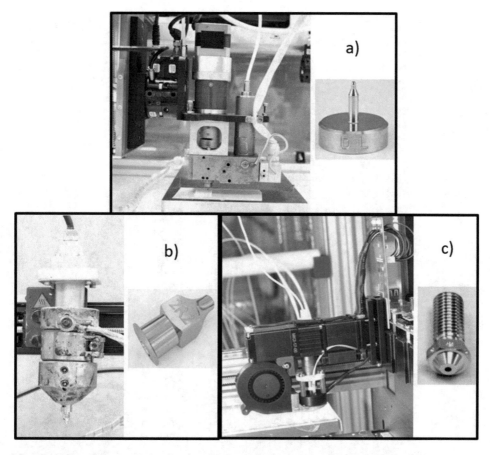

Fig. 4 Detailed view of the extrusion head and the needle/nozzle of the three devices of choice. (**a**) SysEng BioScaffolder and DL Technology G22 encapsulation needle, (**b**) EnvisionTEC Bioplotter and stainless steel Luer Lock G21 hypodermic needle, (**c**) Hyrel 3D and 500 μm nozzle

Table 1
List of the printing parameters used for each technique. Parameters not part of the technique or that could not be programmed are classified as non-applicable (N.A.)

Parameter	FDM	3DF	Bioextrusion
Needle ID (μm)	500	514	413
Temperature (°C)	165	195	195
N_2 Pressure (bar)	N.A.	5.5	4.2
Deposition speed (mm/min)	900	1000	600
Strand distance (mm)	N.A.	1.75	1.5
Infill (%)	60	N.A.	N.A.
RPM	N.A.	N.A.	60
Rollers speed (RPM)	400	N.A.	N.A.

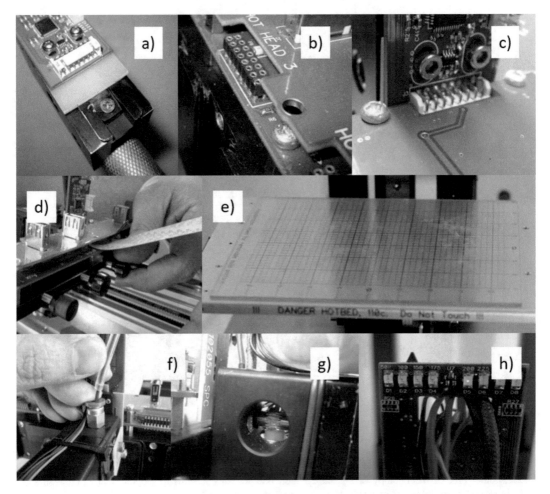

Fig. 5 Detailed view of the components for the head assembling procedure with Hyrel 3D: (**a**) head gib and connector, (**b**) socket for the head connector on the printer, (**c**) proper connection between the head and the printer, (**d**) tightening of the thumbscrew, (**e**) detail of the building platform where to apply the double-sided tape, (**f**) loading of the filament into the blue hose, (**g**) alignment of the filament with the motor's teeth, (**h**) rocker switch for filament loading

6. Calibrate the Z axis (Fig. 6): bring the build surface to the nozzle and save as zero height the position where the tape just starts to pinch.

7. Insert the filament in the blue hose (Fig. 5f) and align it with the motor's teeth (Fig. 5g).

8. Activate the rocker switch to load the filament (Fig. 5h) and lock the filament guide tube.

Generate the G-code for the AM system from the previously created .STL file via the same controlling software:

1. Import the STL file of the part to be printed.

2. Run the slicing software and set the following parameters:

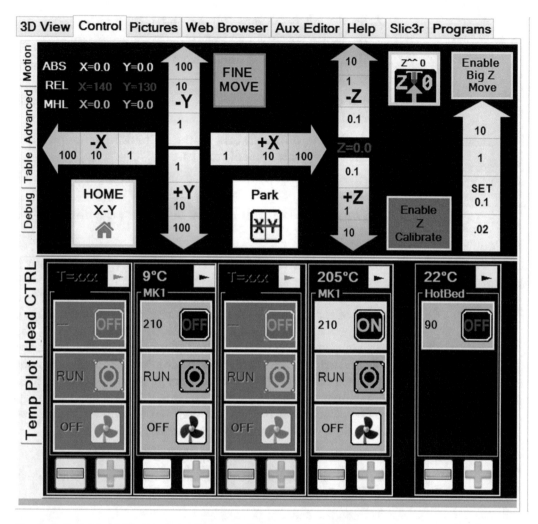

Fig. 6 Control tab of the Repetrel software. In the top part of the window, the commands to move the deposition head can be seen. In the bottom part are the temperature, roller, and fan controllers for each installed head

(a) Layer height: 0.5 mm.

(b) Fill density: 60%.

(c) Fill angle: 90%.

(d) Infill speed: 15 mm/s.

(e) Nozzle internal diameter (ID): 0.5 mm.

(f) *See* **Note 17** for further parameters to be specified.

3. Export the G-code and import it in the controlling software.

By using the same software, the deposition process can now be started:

1. Set the print head temperature to 165 °C and wait for the extruder to heat up.

2. Set the rollers speed to 400 rpm and purge some material until the flow is consistent and smooth.

3. Remove any filament dribble from the extruder nozzle.

4. Start the printing.

5. Pay close attention to the first layers as they are those determining the stability of the printout during the process.

6. At the end of the process, let the scaffolds to cool down and then pour some ethanol at its base.

7. After some time (typically in the order of a couple of minutes), the part can be detached.

8. Dry the scaffold with nitrogen and store vacuum-sealed in a dark and dry place.

3.3.2 3D-Fiber Deposition

Manufacturing with a Bioplotter V1.0 (EnvisionTEC, Fig. 3b) is described as an example of the 3DF process, using a stainless steel Luer lock G21 hypodermic needle (Fig. 4b right) shortened to a length of approximately 15 mm (measured from the beginning of the Luer lock mechanism). The machine works by heating the polymer pellets in a cartridge until the melting point. Then, the polymer melt can be extruded via the needle by the application of a gas pressure. Normally an inert and dry gas (e.g., nitrogen) is used to minimize polymer degradation.

The fabrication process was prepared as follows (brief visual instructions can be found in Supplementary Video 2):

1. Fix the needle on the cartridge (parts shown in Fig. 7a).

2. Slide the stainless steel cartridge inside the heater block (Fig. 7b) and fix it in position.

3. Load the material pellets.

4. Lock the pressure valve (Fig. 7c) onto the cartridge.

5. Start the slicing software.

6. Create a new material dataset with the following parameters (as shown in Fig. 8):

    (a) Feed XY: 1000 mm/min.

    (b) Layer thickness: 0.4 mm.

    (c) Pattern: 90°.

    (d) Strand distance: 1.75 mm.

    (e) Check "Meander."

    (f) *See* **Note 17** for further parameters to be specified.

7. Load the .STL file and select the material dataset just created for all the layers.

8. Generate the .NC-code and save it.

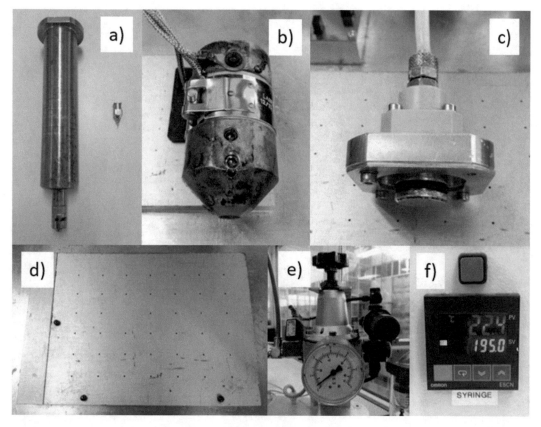

Fig. 7 Detailed view of the components for the head assembling procedure with EnvisionTEC Bioplotter: (**a**) cartridge and needle, (**b**) heater block, (**c**) pressure valve, (**d**) building platform, (**e**) nitrogen line regulator, (**f**) temperature controller

9. Launch the machine controlling software (a detailed view can be seen in Fig. 9) and load the code.

10. Stick a stripe of double-sided tape on the building platform (Fig. 7d).

11. Calibrate the Z axis by moving the print head to the desired printing origin and set this as zero in the software.

12. Adjust the pressure on the compressed nitrogen line regulator (Fig. 7e) to 5.5 bars.

13. Set the temperature to 195 °C on the temperature controller (Fig. 7f) and wait for the cartridge to heat up. *See* **Notes 13–16** for recommendations about the choice of the right temperature for deposition.

14. Purge material until the flow is consistent and smooth.

15. Remove any filament debris from the extruder needle.

16. Press Start on the controlling software to initiate printing.

17. Pay close attention to the first layers as they are those determining the stability of the printout during the process.

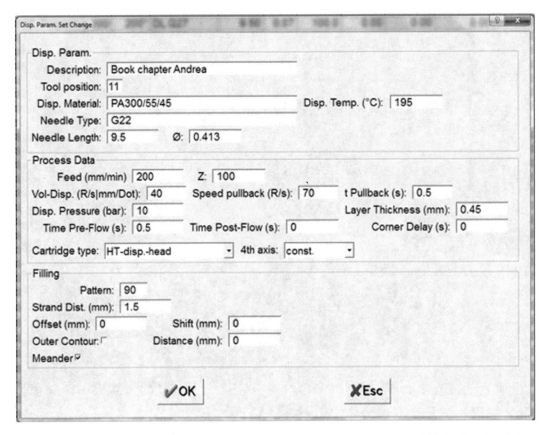

Fig. 8 New dataset window of the PrimCAM slicing software to create a new dataset of fabrication parameters

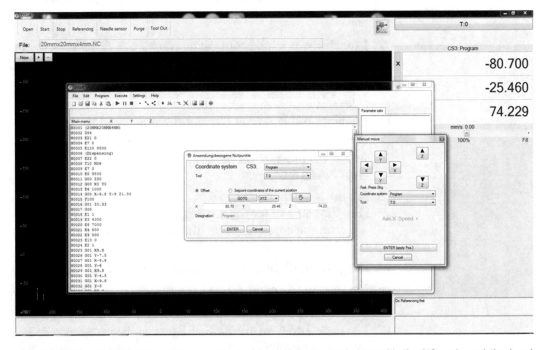

Fig. 9 Details of the DiSoft controlling software. In particular, the window with the NC-code and the head positioning commands can be seen

18. At the end of the process, let the part cool down and then pour some ethanol at its base.

19. After some time (i.e., a couple of minutes), the part can be detached.

20. Dry the scaffold with nitrogen and store vacuum-sealed in a dark and dry place.

3.3.3 Bioextrusion

Manufacturing with a SysEng BioScaffolder (Fig. 3a) equipped with a G22 encapsulation needle (Fig. 4a right) is described as an example of a screw-based AM system. The following steps need to be followed for preparation of the system (brief visual instructions can be found in Supplementary Video 3):

1. Assemble the extrusion head (result of the assembling shown in Fig. 10a).

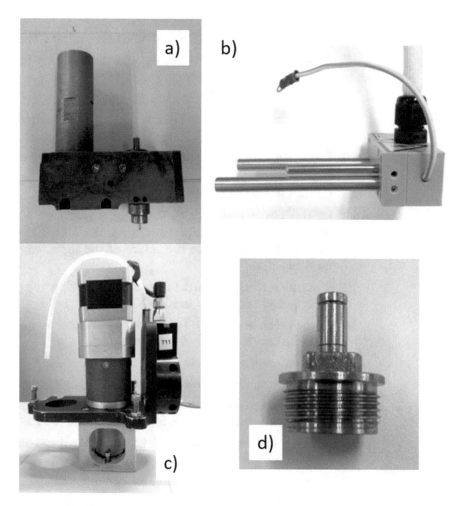

Fig. 10 Detailed view of the components for the head assembling procedure with SysEng BioScaffolder: (**a**) the print head and the reservoir, (**b**) the heating element, (**c**) the motor assembly, (**d**) the nitrogen pressure valve

2. Connect the heater (shown in Fig. 10b) and the motor (Fig. 10c).

3. Load the material and close the reservoir with the nitrogen valve (Fig. 10d).

4. Mount the extrusion head in place and turn the machine on.

5. Start the slicing software.

6. Create a new material dataset with the following parameters:

    (a) Feed XY: 200 mm/min.

    (b) Vol-Disp. (rpm): 100.

    (c) Disp. pressure: 10 bar.

    (d) Layer thickness: 0.21 mm.

    (e) Pattern: 90°.

    (f) Strand distance: 1 mm.

    (g) Check "Meander."

    (h) *See* **Note 17** for further parameters to be specified.

7. Load the .STL file and select the material dataset just created for all the layers.

8. Generate the .NC-code and save it.

9. Launch the machine controlling software (Fig. 9) and load the .NC-code.

10. Stick a stripe of double-sided tape on the building platform.

11. Set the zero point by moving the extrusion head to the desired manufacturing origin.

12. Set the temperature to 195 °C on the temperature controller and wait for the cartridge to heat up. *See* **Notes 13–16** for recommendations about the choice of the right temperature for deposition.

13. After the cartridge has reached the setpoint, allow the polymer to properly melt for around 15–20 min.

14. In the machine controlling software, click the purge button until the flow is consistent and smooth.

15. Remove any filament debris from the extruder needle.

16. Press Start on the controlling software.

17. Pay close attention to the first layers as they are those determining the stability of the printout during the process.

18. At the end of the process, let the part cool down and then pour some ethanol at its base.

19. After some time (i.e., a couple of minutes), the part can be detached.

20. Dry the scaffold with nitrogen and store it vacuum-sealed in a dark and dry place.

3.4   SEM Analysis

To analyze the scaffold under scanning electron microscope, the steps below should be followed (described here for a Philips XL30 SEM):

1. Soak the scaffold in liquid nitrogen for 5 min.

2. Cut the scaffold with a razor blade along the vertical axis.

3. Apply a carbon conductive tape to the sample holders, and stick the samples on them with the section of interest facing upward.

4. Load the scaffolds on the rotating platform inside the gold coating chamber of the sputter coating apparatus in use. Check **Note 18** for the importance of a proper coating.

5. Turn on the machine and apply vacuum.

6. Flush four times with argon.

7. Sputter coat with gold for 100 s.

The samples are then ready to be loaded in the SEM and imaged:

1. Vent the chamber and open it.

2. Load the sample holders onto the platform and close the chamber.

3. Apply vacuum. Follow the recommendation in **Note 19** for proper imaging.

4. Switch on the electron beam and set it at 10 kV.

5. Focus at a low magnification and then go to the desired magnification and focus again. Adjust stigmatism if required.

6. Set the working distance at the characteristic value for the microscope in use and adjust the focus again. Repeat this step until the actual working distance value is as close as possible to the characteristic one.

7. Image the cross section of the scaffolds to have a clear view of the filament diameter and of the lateral porosity. To avoid damages to the scaffold, *see* **Note 20**.

8. Measure the filaments diameter, the filament distance, and the layer thickness in the SEM image to characterize the scaffold.

4   Notes

1. To avoid unreliable data during DSC measurements and that the pan sticks in the machine, ensure that the scale, the tweezers, the pans, and the lids are clean.

2. The sample should be cut so that to avoid irregularities in shape. If the contact area between the sample and the pan is sufficiently uniform, the heat transmission will be regular.

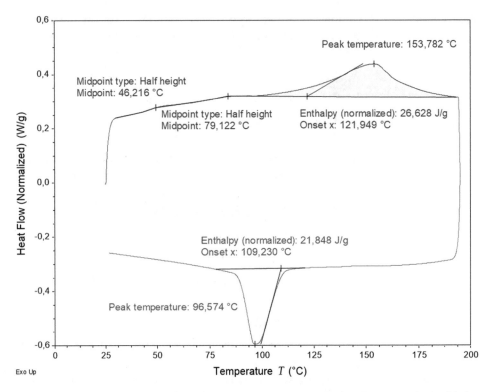

Fig. 11 DSC thermogram of 300PEOT55PBT45 measured at a heating/cooling rate of 10 °C/min under nitrogen atmosphere. Positive heat flow corresponds to endothermic processes taking place in the sample. The blue line (heating cycle) shows two glass transitions (typical for such a diblock copolymer) and one melting peak. During cooling (red line), the material crystallized and reached a crystalline content nearly as high as in its initial state (as can be seen when comparing the enthalpy values)

Consequently, the baseline will be flat and with no sudden steps. For example, a good smooth scan is demonstrated in Fig. 11.

3. In case of comparative studies, ensure equal sample weights to minimize heat transfer effects and consequential thermal lags.

4. To avoid material degradation during the measurement, the final temperature of the heating cycle should not be too high (about 50 °C above the expected peak melting temperature), and the isothermal time should be limited. If the melting temperature is ambiguous, perform thermogravimetric analysis (TGA) first to identify the onset of thermal degradation, which negatively affects an adequate DSC measurement.

5. To eliminate the effect of thermal history, i.e., processing routes, on the melting temperature (T_m), it might be necessary to reheat the sample in a second ramp to the maximum temperature (T_m + 30 °C). The melting temperature thus obtained is intrinsic to the material of interest.

6. To avoid unnecessary additional thermal stress on the sample during rheometry, perform sample loading as fast as possible. In case of thermally unstable materials, one can also pre-shape a disc via compression molding or preferred solution casting excluding any negative thermal effects prior to measurement.

7. Before setting the measurement gap, wait for the temperature to stabilize. During reheating after trimming, the sample viscosity could drop again, and material flow out of the plate might take place.

8. If the measuring gap is too big and the sample exhibits low viscosity, there might be outflow of material resulting in a sudden drop in viscosity. When in a melt state, regular temperature ramps should show a regular increasing trend of complex viscosity with decreasing temperature, as shown in Fig. 12.

9. Outflow could take place also when running frequency sweeps on a low-viscosity samples at very low and very high frequencies. This is due, respectively, to a stronger viscous component and to inertia. If the chosen frequency range is appropriate, the (viscosity vs. frequency) curves should be flat when on the Newtonian plateau and gradually decreasing at higher rates, as can be seen in Fig. 13.

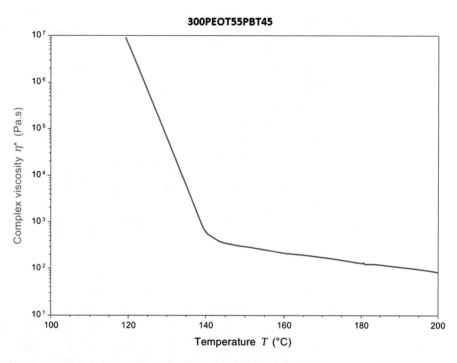

Fig. 12 Temperature dependence of complex viscosity of 300PEOT55PBT45, measured at 1% of applied strain and 1 rad/s, in cooling. The sudden increase in viscosity at around 140 °C is due to the onset of the crystallization process, highly dependent on the cooling

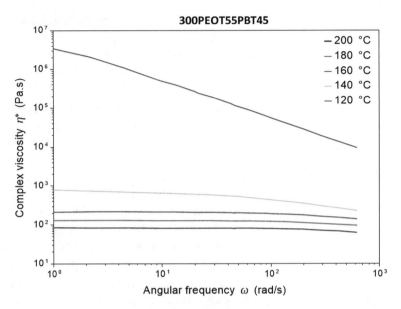

Fig. 13 Frequency dependence of complex viscosity of 300PET55PBT45 at decreasing temperatures, measured at 1% applied strain. It can be seen that the Newtonian plateau shifts to lower frequencies with decreasing temperature. In particular, the crystallization is already complete at 120 °C and the Newtonian plateau cannot be seen

10. Tighten the screws during the assembly of the extrusion heads to avoid polymer leakages, but not too strongly in order to avoid mechanical damage.

11. When setting the origin of the vertical axis, move the print head slowly to avoid crashing the needle against the printing stage.

12. Getting the desired scaffold morphology is the result of a balance between temperature, extrusion driving forces (rollers speed, pressure, and the combination of pressure/screw speed for FDM, 3DF, and bioextrusion, respectively), and deposition speed. These parameters are material and machine dependent. The right parameters to obtain the desired morphology need to be determined experimentally through a feedback loop of fabrication and microscopy analysis. In Fig. 14, it is possible to see the parts fabricated with the set of parameters that led to fiber thicknesses as close as possible to the nozzle internal diameters, as later confirmed by SEM analysis.

13. The higher the temperature and the driving forces, the higher the flow rate. This means that more material is deposited per length, practically resulting in thicker fibers and sagging. As a result, the lateral porosity may be compromised. To avoid this scenario, the deposition speed should be increased accordingly. Examples of properly printed scaffolds and their average fiber

Fig. 14 Stereomicroscope images of scaffolds fabricated with the final set of parameters for Bioplotter (**a**), BioScaffolder (**b**), and Hyrel 3D (**c**). The morphology looks regular with no surface defects and quick measurements confirmed that the fiber diameters are close to the desired ones. Additionally, some free thin filaments can be seen in the background of images (**a** and **b**). These sometimes form as the printing nozzle finishes the deposition of one filament and moves to the next point as explained later

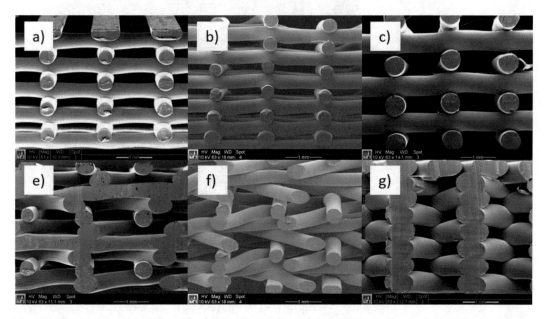

Fig. 15 SEM images of scaffolds fabricated with Bioplotter (**a, e**), BioScaffolder (**b, f**), and Hyrel 3D (**c, g**), following a 0–90 pattern (**a, b, c**), 0–45 (**e, f**), and 45–45 (**g**). The pictures clearly show a regular structure with open lateral pores and no sagging

dimensions are shown in Fig. 15 and Table 2, respectively. In Fig. 16 the result of slow printing speed can be seen instead.

14. Conversely, by reducing the temperature or the driving forces, the flow rate decreases. This could lead to thinner fibers or, in extreme cases, to broken fibers, but also to limited interaction between filaments of subsequent layers. This last effect is due to the shorter time needed for the filament to solidify and to interact with the previous layer.

15. The DSC measurements are needed to evaluate the temperature range when an unknown material is in the melt state and

Table 2
Measured filament thickness (9 measurements) with SEM and comparison with nominal nozzle internal diameter

Dimensions (μm)	FDM	3DF	Bioextrusion
Nozzle	500	514	413
Filament	505 ± 18	523 ± 19	401 ± 35

The chosen set of parameters for each technique allowed to reach dimensions close to the programmed ones

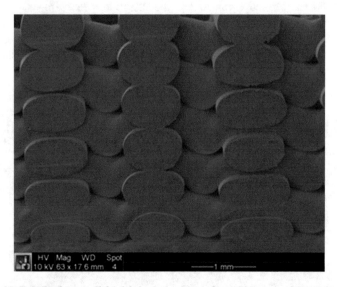

Fig. 16 Detail of a scaffold fabricated with a deposition speed too low with respect to the set flow rate. No lateral porosity can be seen and the cross section of the filaments is highly elliptical

therefore suitable for processing. Depending on the apparatus, semicrystalline thermoplastics are usually processed at temperatures from 10 to 30 °C higher than the melting point. In case of amorphous grades, rheological measurements are needed to give further insights about the material processing window. Rheometry is used to evaluate the material viscosity and therefore how it flows at different temperatures. Increasing the temperature promotes flow but increases the solidification time after deposition, which could cause loss of geometrical accuracy. By reducing the temperature, the risk of degradation decreases, but higher driving forces for deposition are needed.

16. With 3D-fiber deposition and bioextrusion, the duration of the fabrication session should be limited as the material in the reservoir might thermally degrade over time. Lowering the temperature can be an option, but this would influence the flow as well.

Table 3
Comparison of the main characteristics of the three different techniques

Characteristics	3DF	Bioextrusion	FDM
Control on flow rate	Poor with just gas pressure; decent with plunger or piston pressure	Good control, can be quickly switched by changing screw speed or rotation direction, possible also during fabrication if the software includes this functionality	Good control, can be quickly switched by changing roller speed, possible also during fabrication if the software permits it
Thermal degradation	High due to long exposures to high temperatures	High due to long exposures to high temperatures	Low due to short exposures to high temperatures, just before extrusion. However, this advantage can be compromised by the possible requirement of an extra thermal step to make filaments
Ease of use	Complicated in use. Slightly fewer parts to disassemble for cleaning than in screw extrusion and one less parameter to control, i. e., screw speed	Most complicated of the three methods. Involves complicated assembly/disassembly and cleaning as well as the need to regulate extra fabrication parameters	Easy to use. Molten material present only in a small region and can be removed along with the initial extrusion of the next material. To best avoid contaminations though, a more thorough cleaning can be done by disassembling and cleaning parts if the extrusion head permits it
Addition of fillers	Possible, but almost no mixing in molten state	Possible, with slight mixing in the screw	Not possible, unless source filament already includes fillers
Material compatibility	Wide range of materials and forms including pellets and powders	Wide range of materials and forms including pellets and powders	Only materials that can be manufactured into appropriate size filament
Material wastage	Dead volumes lead to material waste, can be minimized by thoughtful design	Dead volumes lead to material waste, can be minimized by thoughtful design	Minimal material waste due to absence of large dead volumes. Material directly passes from the rollers to the manufacturing substrate through a nozzle
Working temperature	High, in order to melt material enough to flow under pressure	High, in order to melt material enough to flow under pressure	Slightly lower temperatures, since the rollers can generate higher pressure on the filament and the melt
Minimum fiber diameter	Dependent on available needle/nozzle and material used (commonly around 150–200 μm)	Dependent on available needle/nozzle and material used (commonly around 150–200 μm)	Dependent on available needle/nozzle and material used (commonly around 150–200 μm)

Fig. 17 Examples of defects due to late flow response of the material (**a**) and excessive flow at the end of the deposition (**b**). The red circle in (**a**) highlights an empty spot where the polymer was not deposited because of delayed flow. In (**b**), thin filaments resulting from the viscoelasticity of the material can be seen. As shown previously, when the nozzle finishes depositing one filament and moves to the next deposition point, some thin polymeric strands could form. This is the result of the viscoelastic nature of polymers, where cohesive forces within the material prevent the fiber from being terminated when the flow is stopped

17. As a consequence of the viscoelastic nature of polymers, it might be necessary to set a pre-flow time or pullback to avoid further fabrication defects. If the response of the polymer is delayed when the printer starts to extrude, some voids might be present in the scaffold. This can be clearly seen in Fig. 17a, where the bottom right corner of the scaffold (where the deposition of each layer started from) is missing. A possible solution is to set a proper pre-flow time. Conversely, if the polymer keeps slightly flowing from the needle even after the end of the deposition, thin undesired filament will be present all over the scaffold. The pullback (only available in bioextrusion) allows applying a backpressure to stop the flow. This defect is shown in Fig. 17b.

18. Be sure to properly sputter coat the sample. As polymers are not highly conductive, uneven or missing coating might affect the quality of the images due to charge accumulation from the incoming electron beam. This can be ensured, for example, by checking that the samples can rotate freely on the rotating platform without interfering with each other. Improper flushing and too short coating time might affect the quality of the coating as well.

19. The SEM analysis should be started only after the chamber is under vacuum or the electron beam might be disturbed by the particles in the atmosphere.

20. Do not keep the beam focused on the same spot for a long time in order to avoid heating from the highly focused electron beam energy.

References

1. Lee JM, Yeong WY (2015) A preliminary model of time-pressure dispensing system for bioprinting based on printing and material parameters. Virtual Phys Prototyp 10:3–8

2. Mota C, Puppi D, Chiellini F, Chiellini E (2015) Additive manufacturing techniques for the production of tissue engineering constructs. J Tissue Eng Regen Med 9:174–190

3. Li MG, Tian XY, Chen XB (2009) A brief review of dispensing-based rapid prototyping techniques in tissue scaffold fabrication: role of modeling on scaffold properties prediction. Biofab 1:32001

4. Chia HN, Wu BM (2015) Recent advances in 3D printing of biomaterials. J Biol Eng 9:4

5. Moroni L, Boland T, Burdick JA, De Maria C, Derby B, Yoo JJ, Vozzi G (2017) Biofabrication: a guide to technology and terminology. Trends Biotechnol. https://doi.org/10.1016/j.tibtech.2017.10.015

6. Taboas JM, Maddox RD, Krebsbach PH, Hollister SJ (2003) Indirect solid free form fabrication of local and global porous, biomimetic and composite 3D polymer-ceramic scaffolds. Biomaterials 24:181–194

7. Youssef A, Hollister SJ, Dalton PD, Amin R, Knowlton S, Hart A (2017) Current and emerging applications of 3D printing in medicine. Biofabrication 9:1–9

8. Poh PSP, Chhaya MP, Wunner FM, De-Juan-Pardo EM, Schilling AF, Schantz JT, van Griensven M, Hutmacher DW (2016) Polylactides in additive biomanufacturing. Adv Drug Deliv Rev 107:228–246

9. Malda J, Visser J, Melchels FP, Jüngst T, Hennink WE, Dhert WJA, Groll J, Hutmacher DW (2013) 25th anniversary article: Engineering hydrogels for biofabrication. Adv Mater 25:5011–5028

10. Moroni L, de Wijn JR, van Blitterswijk CA (2006) 3D fiber-deposited scaffolds for tissue engineering: influence of pores geometry and architecture on dynamic mechanical properties. Biomaterials 27:974–985

11. Hendriks JAA, Moroni L, Riesle J, de Wijn JR, van Blitterswijk CA (2013) The effect of scaffold-cell entrapment capacity and physicochemical properties on cartilage regeneration. Biomaterials 34:4259–4265

12. Nandakumar A, Cruz C, Mentink A, Tahmasebi Birgani Z, Moroni L, Van Blitterswijk C, Habibovic P (2013) Monolithic and assembled polymer-ceramic composites for bone regeneration. Acta Biomater 9:5708–5717

13. Korpela J, Kokkari A, Korhonen H, Malin M, Narhi T, Seppalea J (2013) Biodegradable and bioactive porous scaffold structures prepared using fused deposition modeling. J Biomed Mater Res Part B Appl Biomater 101:610–619

14. Marchioli G, Di Luca A, de Koning E, Engelse M, Van Blitterswijk CA, Karperien M, Van Apeldoorn AA, Moroni L (2016) Hybrid polycaprolactone/alginate scaffolds functionalized with VEGF to promote de novo vessel formation for the transplantation of islets of langerhans. Adv Healthc Mater 5:1606–1616

15. Di Luca A, Longoni A, Criscenti G, Mota C, van Blitterswijk C, Moroni L (2016) Toward mimicking the bone structure: design of novel hierarchical scaffolds with a tailored radial porosity gradient. Biofabrication 8:45007

Chapter 8

Computer-Aided Wet-Spinning

Dario Puppi and Federica Chiellini

Abstract

Computer-aided wet-spinning (CAWS) has emerged in the past few years as a hybrid fabrication technique coupling the advantages of additive manufacturing in controlling the external shape and macroporous structure of biomedical polymeric scaffold with those of wet-spinning in endowing the polymeric matrix with a spread microporosity. This book chapter is aimed at providing a detailed description of the experimental methods developed to fabricate by CAWS polymeric scaffolds with a predefined external shape and size as well as a controlled internal porous structure. The protocol for the preparation of poly(-ε-caprolactone)-based scaffolds with a predefined pore size and geometry will be reported in detail as a reference example that can be followed and simply adapted to fabricate other kinds of scaffold, with a different porous structure or based on different biodegradable polymers, by applying the processing parameters reported in relevant tables included in the text.

Key words Tissue engineering, Scaffold fabrication, Computer-aided wet-spinning, Polymer processing, Biodegradable polymers, Poly(ε-caprolactone)

---

1 Introduction

The combination of additive manufacturing (AM) with other polymer processing approaches is a current hot research topic aimed at the development of biomedical scaffolds with enhanced complexity in terms of integration of structural elements tailored at different length scales. This strategy can be pursued at (1) an assembly level to obtain bi-/multiphasic scaffolds with compartmented architectures, (2) a fabrication level to obtain bimodal scaffolds with fully integrated multi-scale architectures, and (3) a technique level to integrate the working principles of the two processes in a novel hybrid technique for manufacturing scaffolds with a single multifunctional architecture [1]. Successful examples of hybrid technologies development are represented by the integration of AM with solution-electrospinning, melt-electrospinning, freeze-drying, or wet-spinning. In particular, computer-aided wet-spinning (CAWS) has emerged in the past few years as a hybrid AM

Alberto Rainer and Lorenzo Moroni (eds.), *Computer-Aided Tissue Engineering: Methods and Protocols*,
Methods in Molecular Biology, vol. 2147, https://doi.org/10.1007/978-1-0716-0611-7_8,
© Springer Science+Business Media, LLC, part of Springer Nature 2021

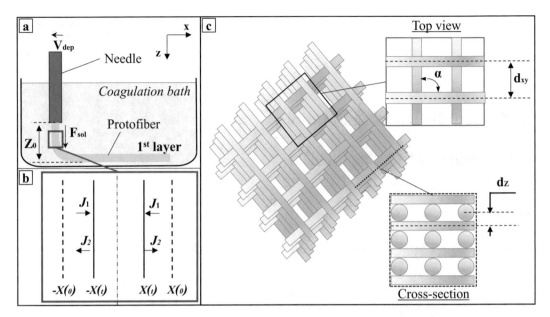

Fig. 1 Scaffold fabrication by CAWS: (**a**) schematic representation of the fiber coagulation process during CAWS showing the needle translating at a resulting deposition velocity (V_{dep}) and at a given initial distance from the collection plane (Z_0) and the polymeric solution extruded at a given feed rate (F_{sol}) directly in a coagulation bath. (**b**) Schematic representation of a detail of the solvent/non-solvent interface where J_1 is the non-solvent flux and J_2 is the solvent flux, and X is the position of the interface at different time points. (**c**) Schematic representation of a designed scaffold architecture showing structural parameters corresponding to given deposition parameters: interfiber deposition distance (d_{xy}), lay-down pattern angle (α), interlayer needle translation (d_Z)

technique suitable for the layered manufacturing of polymeric scaffolds with a predesigned anatomical shape and a controlled dual-scale porous structure [2, 3]. CAWS is based on the computer-controlled deposition of a polymeric solution filament into a coagulation bath (typically a non-solvent of the polymer) to build up 3D scaffolds with a layer-by-layer process. During its extrusion through a needle immersed in the coagulation bath, the solution filament solidifies because of polymer desolvation caused by solvent/non-solvent exchange that leads to the separation of the initial thermodynamically stable solution into two phases with different composition (Fig. 1). Under a critical composition of the polymeric solution, a polymer-lean phase is dispersed into a continuous polymer-rich phase finally resulting in the formation of a local microporosity in the polymeric matrix. The final dual-scale scaffold architecture is characterized by a fully interconnected network of macropores, with size and geometry determined by the predefined lay-down pattern, and a micro-/nanoporosity that can be tailored by acting on different phase inversion parameters.

The optimization of customized manufacturing protocols has led to the development of a set of layered scaffold prototypes made of different materials, including synthetic polymers, such as poly (ε-caprolactone) (PCL) [4, 5], a three-arm star PCL (∗PCL) [6–8], and a poly(ethylene oxide terephthalate)/poly(butylene terephthalate) copolymer [9], a microbial biodegradable polyester, i.e., (poly[(R)-3-hydroxybutyrate-co-(R)-3-hydroxyhexanoate)] (PHBHHx) [10, 11], PHBHHx/PCL blends [12], and novel chitosan/poly(γ-glutamic acid) polyelectrolyte complexes [13, 14]. In addition, an innovative CAWS approach, involving the controlled deposition of the coagulating fiber onto a rotating cylinder, was recently developed to fabricate PCL and PHBHHx samples with a tubular geometry investigated as potential biodegradable stents for small caliber blood vessels treatment [15].

In this book chapter, the protocol to manufacture PCL and hydroxyapatite (HA)-loaded PCL scaffolds by CAWS is reported as a representative example. The procedures for the preparation of PCL solutions, the fabrication of PCL-based scaffolds, and the post-fabrication treatment of the produced samples are described in details. By following the same sequential preparation steps, other kinds of polyester-based scaffolds can be easily fabricated by employing the processing parameters reported in relevant experimental tables included in the chapter.

2   Materials

2.1   Reagents

1. Poly(ε-caprolactone) (PCL, Mw = 80,000 g/mol).
2. Hydroxyapatite (HA) nanoparticles (size <200 nm).
3. Acetone.
4. Absolute ethanol.

2.2   Fabrication Setup

1. An AM machine enabling the deposition of polymeric solutions into a coagulation bath with a predefined 3D pattern and at a controlled feed rate (Fig. 2) (see **Note 1**).

3   Methods

For a given polymer, the scaffold fabrication process requires identifying a set of optimized parameters, including the polymer's molecular weight (Mw) and concentration (C), the solvent/non-solvent system, the needle gauge, the deposition velocity (V_{dep}), the solution feed rate (F_{sol}), the initial distance between the needle tip and the bottom of the deposition beaker (Z_0), the interfiber needle translation (d_{XY}), and the interlayer needle translation (d_Z) (Fig. 1). The experimental steps for the preparation of PCL80$_{0.5}$ scaffolds (Mw = 80,000 g/mol, d_{XY} = 0.5 mm) (Fig. 3) and the corresponding composite PCL80/HA$_{0.5}$ scaffolds are described

Fig. 2 CAWS apparatus: representative pictures of the CAWS apparatus employed for polymeric scaffold fabrication and the layer-by-layer deposition process

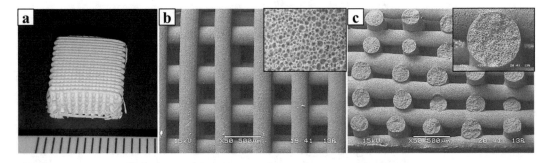

Fig. 3 PCL80$_{0.5}$ scaffolds: (**a**) representative picture of a 3D scaffold (measure unit = 1 mm); representative scanning electron microscopy micrographs of scaffold's (**b**) top view and (**c**) cross-section (insert high magnification images show porosity of the fibers)

in details in the following subsections. The experimental details for the preparation by CAWS of the polymeric mixtures used in the fabrication of a set of polyester-based scaffolds reported in literature and the corresponding optimized fabrication parameters are summarized in Tables 1 and 2, respectively.

Table 1
Experimental details for the preparation of polymeric mixtures used in the fabrication of polyester-based scaffolds by CAWS

Name	Polymer	Solvent	C (w/v %)	Dissolution conditions	Additives	Ref.
PCL80$_{0.5}$	PCL (Capa 6800, Mw = 80,000 g/mol)	Acetone	8	35 °C, 3 h under gentle stirring	–	[4]
PCL80/ HA$_{0.5}$	PCL (Capa 6800, Mw = 80,000 g/mol)	Acetone	8	35 °C, 3 h under gentle stirring	HA (PCL:HA weight ratio 4:1), 2 h under vigorous stirring	[4]
PCL80$_1$	PCL (Capa 6800, Mw = 80,000 g/mol)	Acetone	10	35 °C, 3 h under gentle stirring	–	[4]
PCL80/ HA$_1$	PCL (Capa 6800, Mw = 80,000 g/mol)	Acetone	10	35 °C, 3 h under gentle stirring	HA (PCL:HA weight ratio 4:1), 2 h under vigorous stirring	[4]
PCL50	PCL (Capa 6500, Mw = 50,000 g/mol)	Acetone	20	35 °C, 3 h under gentle stirring	–	[5]
PCL50/ HA	PCL (Capa 6500, Mw = 50,000 g/mol)	Acetone	20	35 °C, 3 h under gentle stirring	HA (PCL:HA weight ratio 4:1), 2 h under vigorous stirring	[5]
∗PCL	∗PCL (Mw = 189,000 g/mol)	Acetone	20–30	35 °C, 3 h under gentle stirring	–	[6]
∗PCL/ HA	∗PCL (Mw = 189,000 g/mol)	Acetone	30	35 °C, 3 h under gentle stirring	HA (PCL:HA weight ratio 4:1), 2 h under vigorous stirring	[6]
PHBHHx	PHBHHx (12 mol% HHx, Mw = 130,000 g/mol)	Chloroform	25	30 °C, 2 h under gentle stirring	–	[10]

PCL poly(ε-caprolactone); *HA* hydroxyapatite; ∗*PCL* star poly(ε-caprolactone); *PHBHHX* poly[(R)-3-hydroxybuty-rate-co-(R)-3-hydroxyhexanoate)]; *C* polymer concentration; *T* temperature; *t* time

Table 2
Optimized parameters for the fabrication of polyester-based scaffolds by CAWS

Sample[a]	d_{XY} (mm)	d_Z (mm)	V_{dep} (mm/min)	F_{sol} (mL/h)	Z_0 (mm)	Needle gauge (inner diameter)	Ref.
PCL80$_1$ PCL80/HA$_1$	1.0[b]	0.2	300	1.4	5.0	23 (0.34 mm)	[4]
PCL80$_{0.5}$ PCL80/ HA$_{0.5}$	0.5	0.2	170	1.0	1.0	23 (0.34 mm)	[4]
PCL50 PCL50/HA	0.5, 1.0, 1.5, 2.0	0.1	240	1.0	2.0	22 (0.41 mm)	[5]
*PCL *PCL/HA	0.5	0.2	240	1.0	1.0	23 (0.34 mm)	[6]
PHBHHx	1.0, 0.5 (0.2)[c]	0.1	600	0.5 (0.3)[c]	2.0	22 (0.41 mm)	[10]

d_{XY} interfiber needle translation; d_Z interlayer needle translation; V_{dep} resulting needle translation velocity; F_{sol} solution feed rate; Z_0 initial needle tip-deposition plane distance

[a]In all cases, ethanol was used as non-solvent

[b]0.5 mm staggered fiber spacing between successive layers with the same fiber orientation was adopted (*see* **Note 3**)

[c]In the case of $d_{xy} = 0.2$ mm, the optimized value of F_{sol} was 0.3 mL/h

3.1 Polymeric Mixtures Preparation

1. For the fabrication of PCL80$_{0.5}$ scaffolds, add 400 mg of PCL to acetone in a 10 mL glass flask, and left the mixture at 35 ° C under gentle magnetic stirring for 3 h in order to prepare 5 mL of a homogenous solution at 8% w/v concentration (*see* **Note 2**).

2. For the fabrication of PCL80/HA$_{0.5}$ composite scaffolds, add 80 mg of HA to a PCL solution prepared as described in **step 1**, and left the suspension at 35 ° C under vigorous stirring for 2 h to achieve a homogeneous dispersion.

3.2 Scaffold Fabrication

The sequential steps for the fabrication of PCL80$_{0.5}$ and PCL80/HA$_{0.5}$ scaffolds are following described.

1. Set the optimized lay-down parameters as following reported: (1) $d_{xy} = 0.5$ mm; (2) $V_{dep} = 170$ mm/min; (3) $d_z = 0.2$ mm (*see* **Notes 3** and **4**).

2. Place the PCL solution or PCL/HA mixture in a plastic syringe (10 mL volume), and mount a blunt tip stainless steel needle (gauge 23) on it (*see* **Notes 5** and **6**).

3. Place the syringe in the programmable syringe pump mounted on the AM machine (*see* **Note 7**).

4. Fix a clean glass beaker on the deposition platform (*see* **Note 8**).

5. Set a Z_0 of 2 mm, and set the X_0 and Y_0 coordinates properly, depending on the instrument, fabrication software, and beaker size used (*see* **Note 9**).

6. Fill the coagulation bath beaker with ethanol such that the liquid level is at least 15 mm (*see* **Notes 10** and **11**).

7. Move the needle to X, Y coordinates out of the deposition area, and set a Z coordinate value for the needle tip position of at least 10 mm (*see* **Note 12**).

8. Start solution feeding at a flow rate of 4 mL/h (*see* **Note 13**).

9. When the extruded filament has a continuous and homogenous morphology, decrease the flow rate to 1 mL/h and wait at least 10 s, checking that the filament flow is eventually stabilized (*see* **Note 14**).

10. Start the computer-controlled motion of the needle and the construction platform (*see* **Notes 15–17**).

3.3 Post-fabrication Treatment

1. Remove the samples from the coagulation bath, and place them under a fume hood for 48 h (*see* **Note 18**).

2. Place the samples in a vacuum chamber for at least 16 h.

3. Store the samples in a desiccator.

4 Notes

1. The rapid prototyping system (MDX-40A; ROLAND DG Mid Europe Srl, Ancona, Italy) modified in-house by replacing the milling head unit with a syringe pump system (NE-1000; New Era Pump Systems Inc., Wantagh, NY, USA) and fixing to the construction platform a beaker containing the ethanol as coagulation bath [6] can be considered as a reference CAWS equipment (Fig. 2). In this case, the lay-down pattern for scaffold production is either calculated using an algorithm developed in MATLAB software (The MathWorks, Inc., Natick, MA, USA) or directly written in G-code and then uploaded into the equipment through the software VPanel for MDX-40A.

2. The prepared PCL solution is stable at room temperature. If the temperature drops appreciably (e.g., during the night), the solution can undergo gelation.

3. In order to fabricate PCL scaffolds with a fiber-to-fiber distance of 1 mm ($PCL80_1$ and $PCL80/HA_1$), 0.5 mm staggered fiber spacing between successive layers with the same fiber orientation is necessary [4]; in addition, in this case C is 10% w/v, V_{dep} 300 mm/min, and F_{sol} 1.4 mL/h (Tables 1 and 2).

4. By employing these optimized parameters, scaffolds composed by different numbers of layers (ranging from 12 to 64) have been produced.

5. When chloroform is used as solvent (e.g., in the case of PHBHHx scaffold fabrication), a glass syringe should be used.

6. Be sure that no air bubbles are present in the syringe in order to have a better control over polymeric solution flow rate during scaffold fabrication. An effective way to place the solution inside the syringe is removing the plunger and mounting a Luer stop cap on the syringe in order to avoid the solution flowing out. After the solution is poured into the syringe under a fume hood, insert only partially the plunger, remove the Luer stop cap, and eliminate the air inside the syringe by pushing gently on the plunger. Finally mount the blunt tip needle on the syringe.

7. Be sure that no solidified polymer is formed where the solution can come in contact with air (e.g., at the tip of the syringe before coupling it to the blunt tip needle or at the tip of the needle while fixing the syringe to the pump due to an accidental pressure on the plunger) in order to not compromise the feed rate control. In case of polymer solution leaking during syringe mounting on the pump, clean the needle tip accurately with solvent-soaked paper.

8. You can use bi-adhesive tape to fix the beaker to the construction platform. The use of a Pyrex glass beaker is recommended to avoid sample detachment during fabrication. In the case of polymers showing high shrinkage degree during coagulation, warming the coagulation bath of few degrees Celsius (e.g., up to 25/26 °C) can be an effective way to avoid sample detachment.

9. It is advised to do the coordinates zeroing before filling the beaker with the non-solvent in order to have a better visual control of the actual needle tip position.

10. In all fabrication processes reported in Table 2, ethanol was employed as a low toxic non-solvent suitable for polymer coagulation; when water was tested as a non-solvent, a homogeneous solution filament was not formed.

11. Although the volumes of solvent and non-solvent employed and evaporated during the process are low, it is recommended to perform the fabrication process under a fume hood.

12. Before starting the fabrication of the first layer, keep the needle tip at a distance from the bottom of the beaker in order to ease the observation of the extruded filament (*see* **Note 13**).

13. After the pump is started, small bubbles exiting from the needle tip should be seen due to air inside the needle. Since

the extruded solution flow can be unsteady during the process start-up, it is necessary to keep the flow rate at a value higher than the optimized one until a homogenous solution filament continuously flowing to the bottom of the beaker is observed.

14. If small solidified polymeric filament fragments or particles attached to the needle tip are formed, remove them with the help of a spatula or a needle before starting the process since they can compromise the fiber deposition process.

15. The reported fabrication studies were optimized in a climatized room at a temperature of 23/24 °C. If the temperature is not controlled, particularly if it is too low, inhomogeneous extrusion as well as sample detachment from the glass surface can occur. In such a case, the use of a local heating source (e.g., an infrared lamp) is recommended to keep the local temperature and humidity conditions at constant values.

16. Be sure that at the end of the deposition process, the extruding needle will stop a few mm away the deposition area (whether it is possible, through a proper G-code file writing). This will assure that no further material is deposited onto the fabricated sample compromising its morphology.

17. Change the coagulation bath with fresh ethanol whenever it appears cloudy. Depending on the thickness of your sample, replace the non-solvent volume every one to four samples. If some polymer residues are present on the bottom of the beaker, be sure to remove them and carefully clean the beaker if necessary.

18. In some cases, it could be necessary to keep the sample into the coagulation bath (for, e.g., 30 min) before collecting it, to allow it to undergo complete solidification.

References

1. Giannitelli SM, Mozetic P, Trombetta M, Rainer A (2015) Combined additive manufacturing approaches in tissue engineering. Acta Biomater 24:1–11

2. Puppi D, Zhang X, Yang L, Chiellini F, Sun X, Chiellini E (2014) Nano/microfibrous polymeric constructs loaded with bioactive agents and designed for tissue engineering applications: a review. J Biomed Mater Res B Appl Biomater 102(7):1562–1579

3. Puppi D, Chiellini F (2017) Wet-spinning of biomedical polymers: from single fibers production to additive manufacturing of 3D scaffolds. Polym Int 66(12):1690–1696

4. Puppi D, Mota C, Gazzarri M, Dinucci D, Gloria A, Myrzabekova M, Ambrosio L, Chiellini F (2012) Additive manufacturing of wet-spun polymeric scaffolds for bone tissue engineering. Biomed Microdevices 14 (6):1115–1127

5. Puppi D, Migone C, Grassi L, Pirosa A, Maisetta G, Batoni G, Chiellini F (2016) Integrated three-dimensional fiber/hydrogel biphasic scaffolds for periodontal bone tissue engineering. Polym Int 65(6):631–640

6. Mota C, Puppi D, Dinucci D, Gazzarri M, Chiellini F (2013) Additive manufacturing of star poly(ε-caprolactone) wet-spun scaffolds for bone tissue engineering applications. J Bioact Compat Polym 28(4):320–340

7. Puppi D, Piras AM, Pirosa A, Sandreschi S, Chiellini F (2016) Levofloxacin-loaded star poly(ε-caprolactone) scaffolds by additive

manufacturing. J Mater Sci Mater Med 27 (3):44

8. Dini F, Barsotti G, Puppi D, Coli A, Briganti A, Giannessi E, Miragliotta V, Mota C, Pirosa A, Stornelli MR, Gabellieri P, Carlucci F, Chiellini F (2016) Tailored star poly (ε-caprolactone) wet-spun scaffolds for in vivo regeneration of long bone critical size defects. J Bioact Compat Polym 31(1):15–30

9. Neves SC, Mota C, Longoni A, Barrias CC, Granja PL, Moroni L (2016) Additive manufactured polymeric 3D scaffolds with tailored surface topography influence mesenchymal stromal cells activity. Biofabrication 8 (2):025012

10. Mota C, Wang SY, Puppi D, Gazzarri M, Migone C, Chiellini F, Chen GQ, Chiellini E (2017) Additive manufacturing of poly[(R)-3-hydroxybutyrate-co-(R)-3-hydroxyhexanoate] scaffolds for engineered bone development. J Tissue Eng Regen Med 11(1):175–186

11. Puppi D, Pirosa A, Morelli A, Chiellini F (2018) Design, fabrication and characterization of tailored poly[(R)-3-hydroxybutyrate-co-(R)-3-hydroxyexanoate] scaffolds by

Computer-aided Wet-spinning. Rapid Prototyp J 24(1):1–8

12. Puppi D, Morelli A, Chiellini F (2017) Additive Manufacturing of Poly(3-hydroxybutyrate-co-3-hydroxyhexanoate)/poly(-ε-caprolactone) Blend Scaffolds for Tissue Engineering. Bioengineering 4(2):49

13. Puppi D, Migone C, Morelli A, Bartoli C, Gazzarri M, Pasini D, Chiellini F (2016) Microstructured chitosan/poly(γ-glutamic acid) polyelectrolyte complex hydrogels by computer-aided wet-spinning for biomedical three-dimensional scaffolds. J Bioact Compat Polym 31(5):531–549

14. Chiellini F, Puppi D, Piras AM, Morelli A, Bartoli C, Migone C (2016) Modelling of pancreatic ductal adenocarcinoma in vitro with three-dimensional microstructured hydrogels. RSC Adv 6(59):54226–54235

15. Puppi D, Pirosa A, Lupi G, Erba PA, Giachi G, Chiellini F (2017) Design and fabrication of novel polymeric biodegradable stents for small caliber blood vessels by computer-aided wet-spinning. Biomed Mater 12(3):035011

Chapter 9

Production of Scaffolds Using Melt Electrospinning Writing and Cell Seeding

Eleonore C. L. Bolle, Deanna Nicdao, Paul D. Dalton, and Tim R. Dargaville

Abstract

Melt electrospinning writing (MEW) is a solvent-free fabrication method for making polymer fiber scaffolds with features which include large surface area, high porosity, and controlled deposition of the fibers. These scaffolds are ideal for tissue engineering applications. Here we describe how to produce scaffolds made from poly(ε-caprolactone) using MEW and the seeding of primary human-derived dermal fibroblasts to create cell-scaffold constructs. The same methodology could be used with any number of cell types and MEW scaffold designs.

Key words Direct writing, Fibroblast, Melt electrospinning writing, Scaffold, Tissue engineering

1 Introduction

Melt electrospinning writing (MEW) (also referred to as melt electrowriting [1] or melt electrohydrodynamic plotting [2]) is an additive manufacturing technique for creating micro- to nano-polymer fibers arranged in a pattern defined by a digitally controlled moving collector. MEW involves the extrusion of a molten polymer to a charged spinneret where an electrified molten polymer jet is drawn toward a moving grounded or oppositely charged collector plate. As the electrified jet travels from the spinneret to the collector, it cools and crystalizes leaving the solid polymer fiber on the collector. As is the case with direct-writing approaches, the movement of the collector defines the lay-down pattern, and due to the gap between the spinneret and collector (the collector distance), the process can be considered non-direct-contact. A schematic of a MEW device is shown in Fig. 1. For a detailed account of the evolution of MEW and examples of many of the possible applications, the reader is guided to these excellent reviews: [3–6].

There are two important advantages of MEW for the manufacture of tissue engineering scaffolds compared to other scaffold

Alberto Rainer and Lorenzo Moroni (eds.), *Computer-Aided Tissue Engineering: Methods and Protocols*,
Methods in Molecular Biology, vol. 2147, https://doi.org/10.1007/978-1-0716-0611-7_9,
© Springer Science+Business Media, LLC, part of Springer Nature 2021

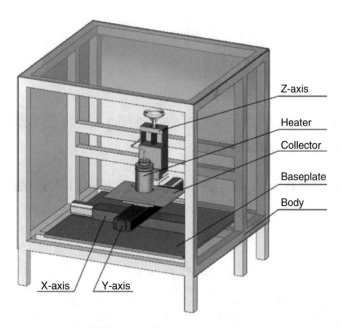

Fig. 1 Schematic of a MEW device. Reproduced under the terms and conditions of the Creative Commons CC BY-NC-ND 3.0 License [7]. Copyright 2016, The Authors, published by De Gruyter

manufacturing techniques. Firstly, it is solvent-free and avoids the use of volatile and toxic solvents. Secondly, the precise placement of MEW fibers produces defined, highly porous structures by leaving spaces between the deposited fibers. These two points contrast with the closely related technique of *solution* electrospinning which is widely researched for the same application and typically results in two-dimensional mats of randomly deposited ultrafine fibers. These solution electrospun meshes essentially have barrier-like properties for anything greater than a few microns in size, as it is difficult to control fiber placement. The high porosity, controlled deposition, and fibrous nature of MEW have made it an attractive technique for preparing tissue engineering scaffolds. Additionally, the scaffolds are ready to use immediately, and the absence of solvent complies with regulatory pathways required to translate a research material into a clinical product [8, 9].

MEW scaffolds can be up to several millimeters thick, thus creating a high surface area, three-dimensional fibrous environment for cells [10]. The seeding of cells can be achieved by simply pipetting a concentrated cell suspension onto the scaffold and allowing the cells to attach. This method has been used to success-fully make scaffold-cell constructs using fibroblasts [10], osteo-blasts (primary human derived and mouse derived) [11, 12], mesothelial cells [12], cardiac progenitor cells [13], T-cells [14], primary human mesenchymal stromal cells from the breast [15], and trabecular bone [16]. So far, MEW has been used to fabricate either flat [10, 17] or tubular [12, 18] scaffolds (Fig. 2). There has

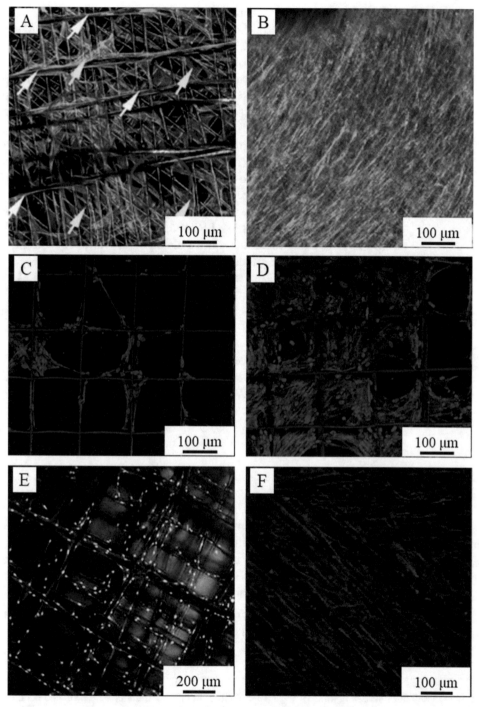

Fig. 2 Examples of various cell types seeded on MEW scaffolds and stained for confocal laser scanning microscopy: Images **a** and **b** show primary human fibroblasts [10] after 7 days (**a**) and 14 days (**b**) of culture. Reproduced with permission [7]. Copyright 2013, IOP Publishing. The samples were stained with FITC-conjugated phalloidin (green) to visualize actin filaments and propidium iodide (red) to visualize cell nuclei. The white arrows indicate fibroblast migrating along the fibers; the yellow arrows indicate fibroblasts spreading across voids. The red circle indicated high fiber density with high fibroblast density and the blue circle low fiber density with low fibroblast density. Images **c** and **d** show primary human mesenchymal stromal cells [16]

also been research into their reinforcement of hydrogels and matrices for superior mechanical properties [19–21]. In addition to their use as scaffolds, MEW fibers have been used as fugitive inks for imparting porosity into hydrogels [22].

One limitation of MEW is the small number of suitable polymers currently tested. To be processed via MEW, candidate polymers must melt at a practical temp (e.g., <200 °C) and have good melt flow properties under the influence of the electrical field (electrorheological properties). To date the only polymers used with MEW are poly(ε-caprolactone) (PCL) [23, 24], poly(hydroxymethylglycolide-co-ε-caprolactone) [13], poly(propylene) [25], poly(L-lactide-co-ε-caprolactone-co-acryloyl carbonate) [26], self-assembling small molecules [27], and poly(2-ethyl-2-oxazoline) [28]. When establishing MEW or working with new polymers, it is important to reduce fiber pulsing—guidelines on identifying a "printability" number and removing fiber pulsing have been published [7, 29]. For tissue engineering applications, using MEW of PCL is particularly attractive as it has a low melting point (60 °C) and good electrorheological properties, has a history of use as implantable medical devices, and is available in medical-grade quality.

In this chapter we describe how to prepare scaffolds from PCL using the MEW process and how these can be seeded with fibroblasts. Because MEW is still an evolving technique, this methods chapter should act as a starting point with the reader encouraged to try their own MEW scaffold designs and different cell types, depending on the type of tissue being mimicked.

2   Materials

2.1   Melt Electrospinning Writing

1. Melt electrospinning machine—we have used an in-house-built machine consisting of high voltage supply (up to +10 kV; low micro-amperage unit), coil electrical heater, computer numerical control (CNC) *x-y* stage controlled with Mach3 CNC controller software, and safety interlocks.

2. Spinneret: 23 gauge, 1″ long Nordson® EFD® 5123 stainless steel precision needle.

Fig. 2 (continued) after 4 days (c) and 10 days (d) of culture. Reproduced under the terms and conditions of the Creative Commons Attribution CC BY 3.0 License [16]. Copyright 2015, The Authors, published by IOP Publishing. The samples were stained with TRITC-conjugated phalloidin (red) to visualize actin filaments and Hoechst 33342 (blue) to visualize cell nuclei. Images e and f show primary human osteoblasts [12] after 14 days (e) and 4 weeks (f) of culture. Reproduced under the terms and conditions of the Creative Commons CC BY License [12]. Copyright 2012, The Authors, published by AIP Publishing. The samples were stained with TRITC-conjugated phalloidin (red) to visualize actin filaments and DAPI (blue) to visualize cell nuclei. e shows an overlay of the fluorescent and transmission light microscopy image

3. 3 cc, 2.88″, clear Nordson® EFD® 7012072 syringe barrel with piston and gas adapter assembly.

4. PURAC® medical grade poly(ε-caprolactone).

2.2 Cell Culture

Here, we have used primary human-derived dermal fibroblasts. Similar work has been performed with 3T3 fibroblasts, mesothelial and mesenchymal stromal cells, keratinocytes, and osteoblasts.

1. Tissue culture flasks.

2. Twelve-well plates.

3. Falcon tubes.

4. Hemocytometer.

5. Fibroblasts growth medium: Complete Dulbecco's Modified Eagle Medium (DMEM). To 500 mL of DMEM, add 5 mL of L-glutamine, 5 mL of penicillin-streptomycin, and 50 mL of fetal calf serum. Store at 4 °C.

6. Sterile phosphate-buffered saline (PBS): without magnesium and calcium.

7. Trypsin-EDTA: 0.05% (Gibco, Thermo Fisher Scientific).

8. Trypan blue.

2.3 Visualization

1. Glutaraldehyde: 3% in cacodylate buffer.

2. 1 M cacodylate buffer.

3. Osmium tetroxide solution: 1% osmium tetroxide in cacodylate buffer.

4. Ethanol/water solutions: 40%, 50%, 70%, 90%, and 100%.

5. Hexamethyldisilazane (HMDS).

6. Paraformaldehyde (PFA): 4% in PBS, supplemented with magnesium and calcium.

7. PBS supplemented with magnesium and calcium: 100 mg of calcium chloride ($CaCl_2$) and 100 mg of magnesium chloride ($MgCl_2 \cdot 6H_2O$) in 1 L of PBS.

8. Permeabilizing buffer: 0.2% Triton X-100 in PBS.

9. Blocking solution: 1% bovine serum albumin (BSA) in PBS.

10. Staining solution: Add 0.8 U/mL FITC-conjugated phalloidin and 5 μg/ML DAPI to blocking solution.

3 Methods

3.1 Fabricating the Scaffolds

1. Tightly pack between 1 and 2 g of the PCL polymer pellets into a 3 cc, 2.88″ syringe barrel, and attach the 23GA, 1″ needle. Place the plastic gas piston over the end of the syringe, and

place the syringe in the oven with the tip facing up for at least 4 h at 100 °C. This allows the PCL pellets to melt and flow to the bottom of the syringe leaving any air to rise to the top.

2. Gently push on the piston to expel the air bubbles through the tip.

3. Load the polymer-filled syringe into the MEW machine, and adjust the distance between the tip of the needle (spinneret) and the collector to 7 mm.

4. Wipe the collector with acetone/ethanol to remove any dust particles/contaminants.

5. On the CNC software, load the G-code (reported in Tables 1 and 2), and move the stage to the desired starting position, and zero the relative *x-y* coordinates (*see* **Note 1**).

6. Turn on the heater. We use a two-stage heater with the barrel of the syringe set to 76 °C and the needle set to 84 °C. Set the gas pressure to 2.1 bar and allow for the formation of a droplet at the tip of the spinneret.

7. After a droplet has formed, turn on the voltage, and slowly increase to 6.6 kV, being aware of any arcing (in which case turn down the voltage) (*see* **Note 2**).

8. Start the execution of the G-code to begin the *x-y* moving. As the jet begins to flow, observe the fiber drag. It may be necessary to vary the collector speed by changing the "f" parameter in the G-code (*see* Tables 1 and 2). Figure 3 shows the effect of collector speed on the fiber drag; a reasonable level of drag is observed in Fig. 3a.

Table 1
G-code command description

G-code	Output description
G17	"xy" plane selection
G21	Program set in mm
G40	Tool radius compensation off—cancels G41 or G42
G49	Tool length offset compensation cancel—cancels G43 or G44
G54	Work coordinate systems, each "tuple" of axis offsets relates to program zero directly to machine zero
G80	Cancel canned cycle
G91	Incremental programming (i.e., relative coordinates)
G94	Feed rate per minute, followed by $F(x)$, where "x" is the units per hour

Table 2
Example G-code. 60 × 60 mm scaffold with fibers in a 90˚ lay-down pattern with spacing of 0.2 mm. The code repeats the pattern 15 times equating to a scaffold 15 layers thick (Fig. 4)

G17 G21 G40 G49 G54 G80 G91 G94 F1500
Description of parameters listed in Table 1
m98 p1
The subprogram "p1" is called, calling the "o1" loop
m98 p2 l15
The subprogram "p2" is called, calling the "o2" loop and repeating it 15 times
m30
Ends the program and rewinds
o1 g1 *x*–5.0 *y*–5.0 f75 m99
Subprogram "o1": the stage moves 5 mm in the negative *x* direction and 5 mm in the negative y direction. f75 is the feed rate of 75 mm/min. This step is to move the stage away from the initial position so the scaffold does not overlap with any fibers deposited before the stage commences moving. m99 ends the subprogram
o2 m98 p3 l150 m98 p4 m98 p5 l150 m98 p6 m99
Subprogram "2" which calls another subprogram "p3" (known as nesting) which loops 150 times and moves on to subprogram "p4." Subprogram "p5" follows and is repeated 150 times, followed by subprogram "p6." m99 ends the subprogram
o3 g01 *x*–60.00 *y*0 f650 g03 *x*0 *y*–0.200 *r*–0.400 f170 g01 *x*60.00 *y*0 f650 g02 *x*0 *y*–0.200 *r*–0.400 f170 m99
Subprogram "3," the stage moves 60 mm in the negative x direction at a feed rate of 650 mm/min. This is where the speed of the stage can be adjusted to optimize the drag. The stage then does a loop with a radius of 0.4 mm, moving counterclockwise and leaving an opening of 0.2 mm. The feed rate here is f170 mm/min. The loops are at a slower speed to allow the fiber to catch up and provide a frame to the scaffold. The stage then moves 60 mm in the positive x direction and does a clockwise loop. m99 ends the subprogram
o4 g1 *x*–60.00 y0 f650 m99
Subprogram "4," the stage moves 60 mm in the negative x direction at a feed rate of 650 mm/mins

(continued)

**Table 2
(continued)**

o5
gl *x*0 *y*60.00 f650
g02 *x*0.200 *y*0 r–0.400 f170
gl *x*0 *y*–60.00 f650
g03 *x*0.200 *y*0 *r*–0.400 f170
m99

Subprogram "5," the stage moves 60 units in the positive y direction at a feed rate of 650 mm/min. The
  stage then does a loop with a radius of 0.4 mm, moving clockwise and leaving an opening of 0.2 mm.
  The feed rate here is f170 mm/min. The stage then moves 60 mm in the negative y direction and does
  a counterclockwise loop. m99 ends the subprogram

o6
gl *x*0 *y*60.00 f650
m99

Subprogram "6," the stage moves 60 mm in the positive y direction at a feed rate of 650 mm/min

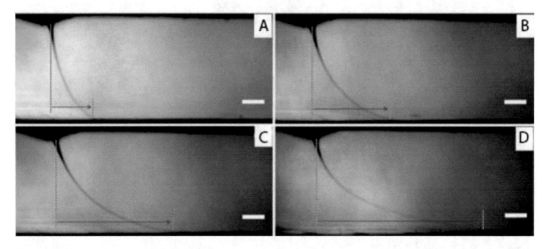

Fig. 3 Effect of collector speed on fiber drag. **a–d** are increasing in collector speed from 500, 1000, 2000,
8000 mm/min, respectively. Scale bars = 1 mm. Reproduced under the terms and conditions of the Creative
Commons CC BY-NC-ND 3.0 License [7]. Copyright 2016, The Authors, published by De Gruyter

9. Once the scaffold printing is complete, switch off voltage fol-
    lowed by gas pressure and temperature.

10. Turn off machine, peel off scaffold, and store safely. Clean any
    remaining sample on the syringe tip and collector with etha-
    nol/acetone and paper towel.

**3.2 Preparing the
Scaffolds for Cell
Culture**

1. Use a biopsy punch or laser cutter to cut the scaffold to the
    desired size.

2. Sterilize the scaffolds by soaking them in 80% ethanol for
    30 min, followed by UV irradiation for 20 min on each side.

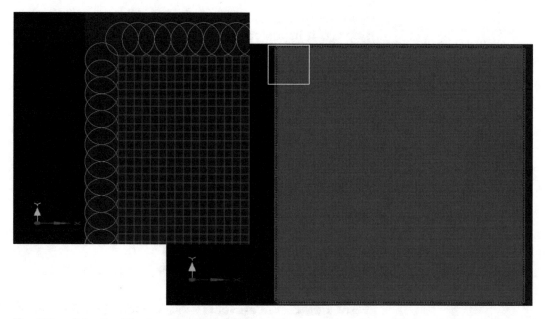

Fig. 4 The print path of the G-code example above. Left: an expansion of the edge area showing the 0.4 mm radius turning loops and the end of each line. Right: the whole 60 × 60 mm design with 0.2 mm between fibers (*see* **Note 3**)

3. Centrifuge the scaffolds in fibroblasts growth medium at $400 \times g$ for 10 min to fully immerse the scaffolds in medium and to minimize the air trapped inside the scaffold.

4. Place the scaffolds in fibroblasts growth medium and keep them in the incubator overnight (*see* **Note 4**).

3.3 Seeding the Scaffolds with Cells

1. Once the cells reach 80% confluence, they can be seeded onto the scaffolds.

2. Remove the culture medium from the flask and rinse cells with 5 mL of sterile PBS.

3. Add 1–2 mL of trypsin to the flask and incubate for 2 min to detach the cells from the flask.

4. Once the cells begin to retract, tap the sides of the flask to fully detach the cells.

5. Add 10 mL of fibroblasts medium to inhibit the trypsin, wash down the flask with the medium, and transfer the cell suspension into a falcon tube.

6. Spin down the cell suspension at $200 \times g$ for 5 min, and resuspend the cell pellet in 10 mL of fibroblasts medium.

7. Perform a cell count to determine the amount of viable cells. Dilute 100 μL of cell suspension in 100 μL of trypan blue, and add 10 μL of the dilution into the hemocytometer.

8. Seed 2×10^4 cells onto the scaffolds, by placing a droplet of cell suspension in the center of the scaffold, using a pipette (*see* **Note 5**). Place the scaffolds in the incubator, at 37 °C in an atmosphere of 5% CO_2 and 95% air, for 2 h, to allow for initial cell attachment (*see* **Note 6**).

9. After the initial cell attachment, add medium until the scaffold is fully submerged, and keep the samples in the incubator. Change medium every 48 h.

Cellular behavior is best visualized using scanning electron microscopy (SEM) or confocal laser scanning microscopy (CLSM).

3.4 Preparing the Samples for SEM

1. All steps are performed in a fume hood at room temperature.

2. Start by fixing the cells in glutaraldehyde for a minimum of 2 h (*see* **Note 7**).

3. Wash the samples in cacodylate buffer for 10 min. Repeat the washes three times.

4. Post-fix the cells in osmium tetroxide solution for 60 min.

5. Rinse the samples in deionized water for 10 min; repeat twice.

6. Dehydrate the cells in a series of ethanol. Start with 40% ethanol for 10 min, followed by 50% ethanol for 10 min. Then place the scaffolds in 70% ethanol for 10 min and 90% ethanol for 10 min, and finish with 100% ethanol for 15 min; repeat the washes twice (*see* **Notes 8** and **9**).

7. Place the scaffolds in HMDS for 30 min; repeat twice (*see* **Note 10**).

8. Allow the scaffolds to fully dry (minimum 1 h) before mounting the scaffolds on stubs.

9. Gold sputter coat the scaffolds at 30 mA for 75 s.

10. Figure 5 shows an example of a MEW scaffold seeded with primary human dermal fibroblasts imaged using a scanning electron microscope.

3.5 Preparing the Scaffolds for CLSM

1. All steps are performed at room temperature.

2. Wash the cells in PBS supplemented with Mg^{2+} and Ca^{2+} twice.

3. Fix the cells in 4% PFA for 20 min.

4. Wash the samples in PBS (*see* **Note 11**).

5. Permeabilize the cells with permeabilizing buffer for 5–10 min.

6. Wash the samples in PBS twice.

7. Incubate the samples in the blocking solution for 5–10 min, to prevent nonspecific binding.

8. Wash the samples in PBS three times.

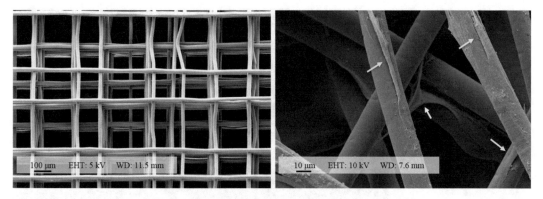

Fig. 5 Left: SEM image of a MEW scaffold. Right: MEW scaffold seeded with primary human-derived dermal fibroblasts. The white arrow shows cells bridging the voids and the yellow arrows show cells migrating along the fibers

9. Incubate the samples in the staining solution for 45 min on a shaker, protected from light.

10. Wash the samples in PBS twice.

11. Store the samples at 4 °C, protected from light until imaging using CLSM. For an example image, *see* Fig. 6.

4   Notes

1. You can simulate the printing time in Mach3 software under the "Tool Path" tab, "Simulate Program Run." The example code provided above takes approximately 15 h to run.

2. Every MEW machine is different, and so optimization of the voltage, temperature, gas pressure, and *x-y* stage speed will be required. The time taken to equilibrate can be quite long as the flow rates are very low, and so it is useful to let the machine run for a few hours with a test scaffold before attempting to print an entire scaffold.

3. The spacing of the fibers will dramatically affect the seeding density. In the above example, the space between fibers is 0.2 mm; spacing greater than this may mean more cells fall through the scaffold during seeding.

4. The scaffolds may be pre-treated to increase cellular adhesion, for example, NaOH etching or plasma treatment.

5. The amount of cells may be adjusted to the size of the scaffold.

6. The cell suspension may be aspirated from the bottom of the well plate every 20 min following initial seeding and reseeded onto the scaffold to maximize cell attachment to the scaffold.

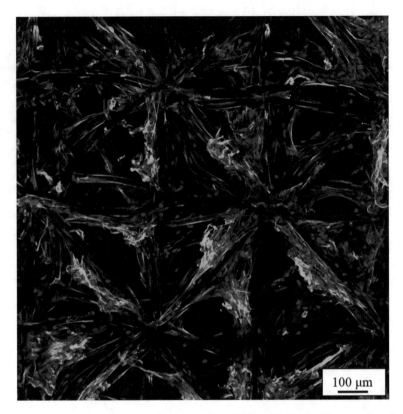

Fig. 6 Primary human dermal fibroblasts on a MEW scaffold with fiber spacing of 0.2 mm. The samples were prepared according to the above protocol and imaged using a CLSM. The cell nuclei were stained with DAPI and the actin filaments with FITC-conjugated phalloidin

7. Glutaraldehyde is a slow fixative and preserves cell morphology well. It may be stored at −20 °C and can be used for up to a week before it will start to crosslink.

8. Don't let the samples dehydrate during SEM sample preparation, as this may affect the cell morphology. The steps are best performed in a well plate. It is not necessary to remove the liquid entirely, as multiple washes are performed.

9. If necessary the dehydration process may be paused at 70% and the samples left at 70% overnight; otherwise the process needs to be backtracked.

10. Depending on the materials used for the scaffolds, the samples may be dried in a critical point dryer following the 100% ethanol incubation.

11. The samples for CLSM may be fixed and stored in PBS at 4 °C and stained at a later time.

References

1. Dalton PD (2017) Melt electrowriting with additive manufacturing principles. Curr Opin Biomed Eng 2:49–57

2. Wei C, Dong JY (2013) Direct fabrication of high-resolution three-dimensional polymeric scaffolds using electrohydrodynamic hot jet plotting. J Micromech Microeng 23:2

3. Hutmacher DW, Dalton PD (2011) Melt electrospinning. Chem Asian J 6:44–56

4. Brown TD, Dalton PD, Hutmacher DW (2016) Melt electrospinning today: an opportune time for an emerging polymer process. Prog Polym Sci 56:116–166

5. Muerza-Cascante ML, Haylock D, Hutmacher DW, Dalton PD (2014) Melt electrospinning and its technologization in tissue engineering. Tissue Eng Part B Rev 21:187–202

6. Dalton PD, Muerza-Cascante ML, Hutmacher DW (2015) Chapter 6. Design and fabrication of scaffolds via melt electrospinning for applications in tissue engineering. 100-120. In: Mitchell G (ed) Electrospinning: principles, practice and possibilities, RSC polymer chemistry series. RSC Publishing, Cambridge

7. Hochleitner G, Youssef A, Hrynevich A, Haigh JN, Jungst T, Groll J, Dalton PD (2016) Fibre pulsing during melt electrospinning writing. BioNanoMaterials 17:159–171

8. Youssef AB, Hollister SJ, Dalton PD (2017) Additive manufacturing of polymer melts for implantable medical devices and scaffolds. Biofabrication 9:012002

9. Pashuck ET, Stevens MM (2012) Designing regenerative biomaterial therapies for the clinic. Sci Transl Med 4:160sr4

10. Farrugia BL, Brown TD, Upton Z, Hutmacher DW, Dalton PD, Dargaville TR (2013) Dermal fibroblast infiltration of poly(epsilon-caprolactone) scaffolds fabricated by melt electrospinning in a direct writing mode. Biofabrication 5:025001

11. Muerza-Cascante ML, Shokoohmand A, Khosrotehrani K, Haylock D, Dalton PD, Hutmacher DW, Loessner D (2017) Endosteal-like extracellular matrix expression on melt electrospun written scaffolds. Acta Biomater 52:145–158

12. Brown TD, Slotosch A, Thibaudeau L, Taubenberger A, Loessner D, Vaquette C, Dalton PD, Hutmacher DW (2012) Design and fabrication of tubular scaffolds via direct writing in a melt electrospinning mode. Biointerphases 7:1–16

13. Castilho M, Feyen D, Flandes-Iparraguirre M, Hochleitner G, Groll J, Doevendans PAF, Vermonden T, Ito K, Sluijter JPG, Malda J (2017) Melt electrospinning writing of poly-hydroxymethylglycolide-co-epsilon-caprolactone-based scaffolds for cardiac tissue engineering. Adv Healthc Mater 6:1700311

14. Delalat B, Harding F, Gundsambuu B, De-Juan-Pardo EM, Wunner FM, Wille M-L, Jasieniak M, Malatesta KAL, Griesser HJ, Simula A, Hutmacher DW, Voelcker NH, Barry SC (2017) 3D printed lattices as an activation and expansion platform for T cell therapy. Biomaterials 140:58–68

15. Weigand A, Boos AM, Tasbihi K, Beier JP, Dalton PD, Schrauder M, Horch RE, Beckmann MW, Strissel PL, Strick R (2016) Selective isolation and characterization of primary cells from normal breast and tumors reveal plasticity of adipose derived stem cells. Breast Cancer Res 18:32

16. Hochleitner G, Jungst T, Brown TD, Hahn K, Moseke C, Jakob F, Dalton PD, Groll J (2015) Additive manufacturing of scaffolds with sub-micron filaments via melt electrospinning writing. Biofabrication 7:035002

17. Brown TD, Edin F, Detta N, Skelton AD, Hutmacher DW, Dalton PD (2014) Melt electrospinning of poly(ε-caprolactone) scaffolds: phenomenological observations associated with collection and direct writing. Mater Sci Eng C Mater Biol Appl 45:698–708

18. Jungst T, Muerza-Cascante ML, Brown TD, Standfest M, Hutmacher DW, Groll J, Dalton PD (2015) Melt electrospinning onto cylinders: effects of rotational velocity and collector diameter on morphology of tubular structures. Polym Int 64:1086–1095

19. Bas O, D'Angella D, Baldwin JG, Castro NJ, Wunner FM, Saidy NT, Kollmannsberger S, Reali A, Rank E, De-Juan-Pardo EM, Hutmacher DW (2017) An integrated design, material, and fabrication platform for engineering biomechanically and biologically functional soft tissues. ACS Appl Mater Interfaces 9:29430–29437

20. Bas O, De-Juan-Pardo EM, Chhaya MP, Wunner FM, Jeon JE, Klein TJ, Hutmacher DW (2015) Enhancing structural integrity of hydrogels by using highly organised melt electrospun fibre constructs. Eur Polym J 72:451–463

21. Visser J, Melchels FPW, Jeon JE, van Bussel EM, Kimpton LS, Byrne HM, Dhert WJA, Dalton PD, Hutmacher DW, Malda J (2015) Reinforcement of hydrogels using three-dimensionally printed microfibres. Nature

Commun 6. https://doi.org/10.1038/ncomms7933

22. Haigh JN, Chuang YM, Farrugia B, Hoogenboom R, Dalton PD, Dargaville TR (2015) Hierarchically structured porous poly (2-oxazoline) hydrogels. Macromol Rapid Commun 37:93–99

23. Dalton PD, Grafahrend D, Klinkhammer K, Klee D, Moller M (2007) Electrospinning of polymer melts: phenomenological observations. Polymer 48:6823–6833

24. Brown TD, Edin F, Detta N, Skelton AD, Hutmacher DW, Dalton PD (2014) Melt electrospinning of poly(epsilon-caprolactone) scaffolds: phenomenological observations associated with collection and direct writing. Mater Sci Eng C Mater Biol Appl 45:698–708

25. Haigh JN, Dargaville TR, Dalton PD (2017) Additive manufacturing with polypropylene microfibers. Mater Sci Eng C 77:883–887

26. Chen F, Hochleitner G, Woodfield T, Groll J, Dalton PD, Amsden BG (2016) Additive manufacturing of a photo-cross-linkable polymer via direct melt electrospinning writing for producing high strength structures. Biomacromolecules 17:208–214

27. Singer JC, Ringk A, Giesa R, Schmidt HW (2015) Melt electrospinning of small molecules. Macromol Mater Eng 300:259–276

28. Hochleitner G, Hummer JF, Luxenhofer R, Groll J (2014) High definition fibrous poly (2-ethyl-2-oxazoline) scaffolds through melt electrospinning writing. Polymer 55:5017–5023

29. Tourlomousis F, Ding H, Kalyon DM, Chang RC (2017) Melt electrospinning writing process guided by a "printability number". J Manuf Sci Eng 139:081004

Chapter 10

Low-Voltage Continuous Electrospinning: A Versatile Protocol for Patterning Nano- and Micro-Scaled Fibers for Cell Interface

Zhaoying Li, Xia Li, and Yan Yan Shery Huang

Abstract

Nano- and micro-scaled fibers have been incorporated in a number of applications in biofabrication and tissue cultures, providing a cell interfacing structure with extracellular matrix-mimicking topography and adhesion sites, and further supporting localized drug release. Here, we describe the low-voltage electrospinning patterning (LEP) protocol, which allows direct and continuous patterning of sub-micron fibers in a controlled fashion. The processable polymers range from protein (e.g., gelatin) to thermoplastic (e.g., polystyrene) polymers, with flexible selections of collecting substrates. The operation voltage for fiber fabrication can be as low as 50 V, which brings the benefits of reducing costs and mild-processing.

Key words Protein patterning, Electrospinning, Direct writing, Nanofiber, Biofabrication, Polymer, Extracellular matrix, Topography, Live cell imaging, Tissue engineering

1 Introduction

One of the main aims of biofabrication is to recreate certain physiological aspects of the extracellular matrix (ECM). Electrospinning is a robust technique for fabricating nano- to microfibers from a wide range of polymer solutions, including both natural and synthetic polymers. The dimensions of these fibers are comparable to the ECM fibril component [1, 2]. Hence electrospun fibers have been used in cell cultures to facilitate structural support and guide cell fate [3–5]. However, there are a few limitations to the conventional electrospinning technique, including high voltage requirement, patterning controllability, and limited protein loading. The high voltage requirement restricts the processability of voltage-sensitive materials [6]. The low patterning controllability is caused by bending instability during the electrospinning process [7]. Modifications to the experimental configuration are required to improve this. In this protocol, a method to achieve controllable patterning

Alberto Rainer and Lorenzo Moroni (eds.), *Computer-Aided Tissue Engineering: Methods and Protocols*,
Methods in Molecular Biology, vol. 2147, https://doi.org/10.1007/978-1-0716-0611-7_10,
© Springer Science+Business Media, LLC, part of Springer Nature 2021

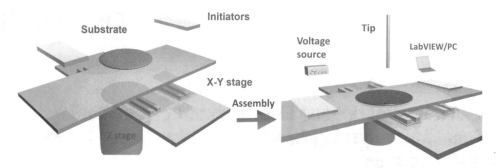

Fig. 1 Schematic diagram of the LEP components and configuration

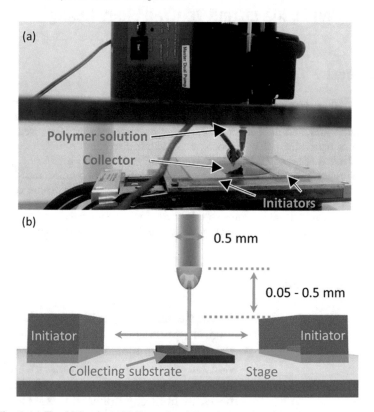

Fig. 2 (a) The LEP setup. (b) Zoom-in of the stage assembly

of polymer fibers is described, of which flexibility is significantly extended compared to conventional electrospinning techniques. The low-voltage electrospinning patterning (LEP) (Figs. 1 and 2) enables precise patterning of a range of polymer solutions on a variety of substrates, with an operation voltage typically at or below 100 V DC [8]. These polymer solutions include, but not limited to, the listed compositions in Table 1. Here, we used three polymer solutions, as examples, to describe the process of LEP. These polymer solutions are: polystyrene (PS) in dimethylforma-mide (DMF), polyethylene oxide (PEO) in water, and a gelatin

Table 1
Example polymer solutions processed using LEP

Solution	Polymer molecular weight (Mw, kDa)	Concentration (wt %)	Conductivity (μS/cm)	Viscosity (Pa·s)	Initiator	Potential application
PEO-water	400	6	99.6	3.1	Insulating	Single bacteria or growth factor encapsulation, patterning and release.
PS-DMF	280	25 30	0.3 0.2	1.4 2.6	Conductive	Topographical guidance for cell attachment and migration.
Gelatin-acids and water	50–100	15	0.9	0.4	Insulating	Combined topography and gelatin-binding receptors for cell culture

aqueous solution. Parallel linear patterns of fibers in the sub-micron dimensions were printed on glass coverslips (Fig. 3) and PDMS films. The patterned fibers can be used as a topographical and chemical guidance for cell attachment and growth [9, 10]. Gelatin fibers patterned on PDMS film were used to culture EA.hy926 cells. The dynamic interactions between cells and the fibers are shown in the live cell images (Fig. 4). Immunostaining was performed to show the cell morphology along a microfiber.

2 Materials

Prepare all solutions using deionized water and analytical grade reagents. Prepare and store all reagents at room temperatures below 25 °C. When disposing waste, strictly follow the waste disposal regulations.

2.1 Example Polymer Solutions

1. PS-DMF: dissolve 0.25 g of PS (Mw = 280 kDa) in 1 g of DMF to obtain a concentration of 25 wt%. Place a magnetic stirrer into the solution and tighten the lid. Leave the solution to stir until homogeneous.

2. PEO-water: dissolve 0.1 g of PEO (Mw = 400 kDa) in 1 g of water to obtain a concentration of 10 wt%. Place a magnetic stirrer into the solution and tighten the lid. Leave the solution to stir until homogeneous.

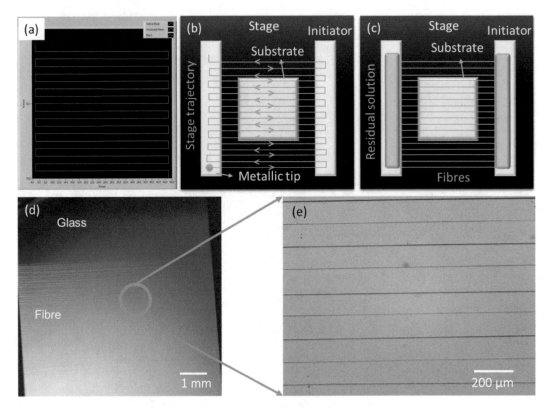

Fig. 3 (**a**) Snapshot of the LabVIEW program, controlling the stage movement. (**b**) The programed path is translated into the stage trajectory, as shown in the green line. (**c**) The fibers are patterned following the stage movement. (**d**) PEO fibers patterned on a glass coverslip. (**e**) Parallel straight fibers imaged under an optical microscope

3. Gelatin: weigh 0.15 g of gelatin powder (porcine skin, Mw = 50–100 kDa). Calculate the amount of solvents required by using the weight ratio of gelatin:water:acetic acid:ethyl acetate = 15:25:36:24. Add the calculated amount of water, acetic acid, and ethyl acetate in sequence. Place a magnetic stirrer into the solution. Tighten the lid and cover the lid with parafilm. Leave the solution to stir until homogeneous.

2.2 LEP

1. DC voltage supplier (up to 5 kV).
2. Crocodile clips connecting to the voltage supplier.
3. Safety switch which ensures the voltage is cut when the switch is off.
4. Syringe pump (World Precision Instruments, AL-1000).
5. X-y moving stage, with a speed of ~100 mm/s, resolution better than 1 μm, and uni-directional repeatability better than 1 μm; and z-stage of resolution better than 1 μm.
6. 1 mL plastic syringe as the polymer solution reservoir.

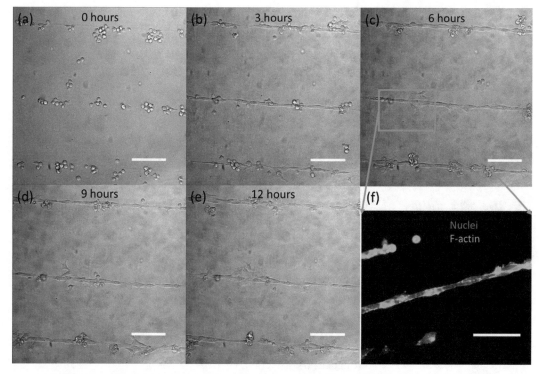

Fig. 4 EA.hy926 cells cultured on gelatin fibers patterned on PDMS film. (**a–e**) Show the live cell imaging at the specific acquisition time. (**f**) Immunostaining of the cells on a gelatin fiber. Scale bars 50 μm

7. Metallic tip (inner diameters 0.3–0.5 mm).

8. Initiators: use glass slides (dimensions approximately 76 × 26 × 1 mm) for PEO-water and gelatin solutions. Cut the silicon wafers (thickness 500–550 μm) into stripes with a width of roughly 10 mm, which can be used for the PS-DMF solutions.

9. Scotch tape.

10. Collecting substrates: a range of materials can be used, such as glass coverslip (dimensions ~22 × 22 × 0.175 mm), silicon wafer (dimensions ~20 × 20 × 0.5 mm), and PDMS film (dimensions ~20 × 20 × 0.1 mm) prepared as described in [9].

11. Software control: a LabVIEW program was constructed to control the x-y stage movement. Figure 3a shows an example to create parallel linear patterns.

2.3 Gelatin Crosslinking Reagents

1. 1-ethyl-3-(dimethyl-aminopropyl) carbodiimide hydrochloride (EDC).

2. N-hydroxyl succinimide (NHS).

3. Mix absolute ethanol with water at a volume ratio of 9:1.

| 2.4 Cell Culture | 1. EA.hy925 cell line (passage 10–20) cultured in T25 flasks. |

2.4 *Cell Culture*

1. EA.hy925 cell line (passage 10–20) cultured in T25 flasks.

2. 5 mL Eppendorf tubes.

3. Round petri-dish (culture area 8.8 cm^2).

4. 70% ethanol.

5. Phosphate-buffered saline (PBS).

6. Trypsin.

7. Cell culture medium: Dulbecco's modified Eagle's medium (DMEM) mixed with 10% fetal bovine serum (FBS) and 1% penicillin-streptomycin.

2.5 *Live cell Imaging*

1. Live cell imaging chamber (temperature and CO$_2$ control).

2. Confocal microscope.

2.6 *Immunostaining*

1. Fixation solution: 4% paraformaldehyde in PBS.

2. Buffer solution: 0.05% Tween-20 in PBS.

3. 0.1% Triton X-100 solution in PBS.

4. Blocking solution: 1% BSA in PBS.

5. Phalloidin.

6. Hoechst (Invitrogen).

3 Methods

Carry out all procedures at room temperature, unless stated otherwise.

3.1 *LEP*

1. Assemble the LEP configuration as shown in Fig. 1.

2. Switch on the stage. Open the LabVIEW program which controls the *x-y* stage movement. Set the stage to move at 100 mm/s in a zig-zag pattern (acceleration 1000 mm/s^2). Test the stage movement without any loading.

3. Load the polymer solution into the syringe. Firmly attach the metallic tip onto the syringe (*see* **Note 1**).

4. Place the syringe in the pump. Set the flow rate of 200 μL/h to match the stage speed. Place the pump in the position where the metallic tip points vertically toward the stage (*see* **Note 2**).

5. Attach one crocodile clip onto the metallic tip. The other crocodile clip (ground) is attached to the stage.

6. Place the collecting substrate onto the stage. The selection of the substrate material depends on the purpose of the experiment. For example, PDMS films and glass coverslips are often used for cell assays. Place two initiators on either side of the collecting substrate, leaving a 0.5 mm gap between the

substrate and each initiator (*see* **Note 3** for the selection of initiators). Use Scotch tape to fix the initiators.

7. Place the collecting platform onto the stage. Check the collecting substrates and initiators are fixed during the stage motion (*see* **Note 4**). Use scotch tape to secure if necessary.

8. Adjust the position of the initiators and the collecting substrate. Tune the stage motion using LabVIEW. Ensure the two initiators move beneath the metallic tip alternatively during stage movement.

9. Manually adjust the height of the z-stage. The distance between the metallic tip and the initiator should be 0.05–0.5 mm. Figure 2 shows the LEP configuration at this point.

10. Start the pump. Wait till a droplet is visible at the end of the metallic tip. Then set the stage into motion. The program controls the stage movement, which is translated into fiber patterns, as shown in Fig. 3a–c.

11. Switch on the voltage (*see* **Note 5**). Applied voltage ranges from 50 V to 700 V for PS-DMF solution, from 50 V to 150 V for PEO solution, and from 100 V to 400 V for gelatin solution (*see* **Note 6** for controlling the fiber morphology).

12. Switch on the safety switch.

13. Wait for the desired pattern to complete. Switch off the voltage and the safety switch.

14. Stop the stage motion using LabVIEW. Lower the stage. Transfer the collecting substrate from the stage to a designated petri-dish (*see* **Note 7**). The fibers should be patterned on the substrate, as shown in Fig. 3d, e.

3.2 Cell Culture

1. Pattern gelatin fibers on a $20 \times 20 \times 0.1$ mm PDMS film using the LEP method as previously described.

2. To crosslink the gelatin fibers, immerse the specimen into a crosslinking solution formed by 25 mM of EDC and 10 mM of NHS in ethanol-water (9:1 v/v) [11]. Store the specimen at 4 °C for 24 h to complete crosslinking.

3. Rinse the specimen with water to remove the crosslinking reagents (*see* **Note 8**).

4. Dry the specimen in a desiccator for 24 h.

5. To sterilize the specimen for cell culture use, immerse it in 70% ethanol for 1 h.

6. Rinse the specimen with PBS for three times. Immerse the specimen in cell culture medium and store at 37 °C.

7. Take a flask of confluent EA.hy926 cells. To resuspend the cells, remove the medium and wash the cells with 2 mL PBS.

Remove the PBS. Immerse the cells in 1 mL trypsin and store at 37 °C for 5 min.

8. Check the cells are in suspension using an optical microscope. Transport the cells into a 5 mL Eppendorf tube. Centrifuge the cells at $180 \times g$ for 5 min to obtain a cell pellet.

9. Remove the excess medium. Add 100 μL cell culture medium and resuspend the cells.

10. Seed the cells onto the gelatin fibers patterned on PDMS to obtain a coverage of 100 cells/mm^2 (*see* **Note 9**). Incubate the cells at 37 °C, 5% CO$_2$.

3.3 Live-Cell Imaging

1. Turn on the live cell imaging chamber. Set the temperature to 37 °C and turn on the CO$_2$ supply. Wait 30 min for the temperature and CO$_2$ level to stabilize.

2. Turn on the confocal microscope and 633 nm laser, acquire in the transmission mode.

3. Place the sample petri-dish into the chamber. Top up the medium if necessary.

4. Adjust the focus of the confocal microscope. Adjust the laser power and digital gain to obtain sharp images. The laser power should be as low as possible to avoid damage to the cells. Find the region of interest.

5. Set the scanning rate to 400 Hz, and image size to 512×512 pixels. Set the acquisition time to 12 h with 10 min intervals.

6. Start imaging. A sequence of live cell images, indicating the interactions between EA.hy926 cells with gelatin fibers patterned on PDMS, is shown in Fig. 4a–e.

3.4 Immuno-fluorescent Staining

1. Immerse the sample in the fixation solution (4% formaldehyde in PBS) for 20 min.

2. Wash the sample twice with the buffer solution (0.05% Tween-20 in PBS).

3. Permeabilize the cells with 0.1% Triton X-100 solution in PBS for 5 min.

4. Wash the sample twice with the buffer solution (0.05% Tween-20 in PBS).

5. Immerse the sample in the blocking solution (1% BSA in PBS) for 30 min.

6. Remove the blocking solution.

7. Add phalloidin (1:400 dilution) and Hoechst (1:10000 dilution) in PBS. Incubate the sample in this solution for 50 min.

8. Wash the sample three times with the buffer solution. Leave the sample in the solution for 5 min during each wash.

9. Store the sample in dark and acquire images using a confocal microscope within 3 days. Figure 4f shows the immunostaining of EA.hy926 cells on a single gelatin fiber.

4   Notes

1. To ensure a steady flow, remove any air bubbles in the syringe. This can be done by withdrawing the piston to the volume limit, and then slowly pushing the piston back to the position where the solution fills the syringe. After the metallic tip is attached to the syringe, push the piston until a droplet of the solution is visible at the end of the tip. Remove the droplet using tissue paper.

2. The stage should be lowered. This is to ensure a safety gap between the stage and the metallic tip to avoid contact. Poking the stage with the metallic tip can damage the precision of the stage movement.

3. The selection of the initiator depends on the electrical conductivity of the polymer solution. Generally, both conductive (silicon wafer) and insulating initiators (glass slides) can be used for solutions with high conductivities, such as PEO-water and gelatin solutions. However, the applied voltage needs to be sufficiently low when conductive initiators are used, in order to avoid electric shorting of the system. Hence insulating initiators are recommended. For solutions with low conductivities, such as PS-DMF solutions, conductive initiators are needed. The polymer solutions and their conductivities are listed in Table 1.

4. Ensure the two initiators are at the same height and are at least 1 mm higher than the collecting substrate. If the collecting substrate is higher than the initiators, solution droplets will be dispensed onto the substrate. If the initiators are at different heights, the solution droplets will not be effectively initiated and the fiber patterning will be hindered.

5. Always set the stage into motion before applying the voltage. This is to avoid short-circuiting, which can cause damage to the voltage supplier and fire hazard.

6. There are three factors contributing to the morphology of patterned fibers: intrinsic polymer solution properties, extrinsic operating parameters, and ambient factors. The intrinsic polymer solution properties include the parameters which reflect the polymer entanglement. The most important parameter is viscosity, which is listed in Table 1 for the solutions processed using LEP. A high viscosity is essential to obtain continuous and uniformly straight fiber patterns. Insufficient viscosity may

lead to discontinuous fiber pattern or beaded defects [8]. The extrinsic operating parameters include voltage and stage speed. An optimal voltage is to be determined for each polymer solution. Insufficient voltage can result in discontinuous fiber pattern, whereas excess voltage can cause increase and fluctuations in fiber diameters. The stage speed should match the polymer dispense rate, so that the droplet is refreshed in time with the initiators. The ambient factors include the temperature, humidity, and air flow. The optimal temperature and humidity should be determined for each polymer solution. For the solutions listed in Table 1, room temperature and humidity are suitable for LEP. The air flow should be at minimum to avoid disruptions to the fiber alignment.

7. Some suspended fibers may be linking the initiators and the collecting substrate. Removing the substrate directly could damage the fiber pattern. A convenient method is to use a razor blade to cut the suspended fibers from the initiators before removing the substrate.

8. Rinse gently to avoid damage to the fiber pattern. Ensure the PDMS is sticking firmly to the bottom of the petri-dish.

9. For cell seeding onto the patterned device, it may be beneficial to remove a small volume of the medium in the petri-dish, so that the medium layer is around 1 mm above the sample surface. Seed the cells and incubate at 37 °C for 12 h to establish cell attachment. Then top up the medium.

Acknowledgment

We would like to thank EPSRC for providing the funding for this study.

References

1. Schiffman JD, Schauer CL (2008) A review: electrospinning of biopolymer nanofibers and their applications. Polym Rev 48:317–352

2. Sill TJ, Recum HAV (2008) Electrospinning: applications in drug delivery and tissue engineering. Biomaterials 29:1989–2006

3. Stitzel J, Liu J, Lee SL, Komura M, Berry J, Soker S, Lim G, Van Dyke M, Czerw R, Yoo JJ et al (2006) Controlled fabrication of a biological vascular substitute. Biomaterials 27:1088–1094

4. Schnell E, Klinkhammer K, Balzer S, Brook G, Klee D, Dalton P, Mey J (2007) Guidance of glial cell migration and axonal growth on electrospun nanofibers of poly-ε-caprolactone and a collagen/poly-ε-caprolactone blend. Biomaterials 19:3012–3025

5. Sahoo S, Ouyang H, Goh JCH, Tay T, Toh S (2006) Characterization of a novel polymeric scaffold for potential application in tendon/ligament tissue engineering. Tissue Eng 12:91–99

6. Zeugolis DI, Khew ST, Yew ES, Ekaputra AK, Tong YW, Yung LYL, Hutmacher DW, Sheppard C, Raghunath M (2008) Electrospinning of pure collagen nano-fibres–just an expensive way to make gelatin? Biomaterials 29:2293–2305

7. Reneker DH, Yarin AL, Fong H, Koombhongse S (2000) Bending instability of

electrically charged liquid jets of polymer solutions in electrospinning. J Appl Phys 87:4531–4547

8. Li X, Li Z, Wang L, Ma G, Meng F, Pritchard RH, Gill EL, Liu Y, Huang YYS (2016) Low-voltage continuous electrospinning patterning. ACS Appl Mater Interfaces 8:32120–32131

9. Xue N, Li X, Bertulli C, Li Z, Patharagulpong A, Sadok A, Huang YYS (2014) Rapid patterning of 1-d collagenous topography as an ECM protein fibril platform for image cytometry. PLoS One 9:e93590

10. Xue N, Bertulli C, Sadok A, Huang YYS (2014) Dynamics of filopodium-like protrusion and endothelial cellular motility on one-dimensional extracellular matrix fibrils. Interface Focus 4:20130060

11. Zhang S, Huang Y, Yang X, Mei F, Ma Q, Chen G, Deng X (2009) Gelatin nanofibrous membrane fabricated by electrospinning of aqueous gelatin solution for guided tissue regeneration. J Biomed Mater Res A 90:671–679

Chapter 11

Direct-Write Deposition of Thermogels

Sara Maria Giannitelli, Valeria Chiono, and Pamela Mozetic

Abstract

The use of biocompatible hydrogels has widely extended the potential of additive manufacturing (AM) in the biomedical field leading to the production of 3D tissue and organ analogs for in vitro and in vivo studies.

In this work, the direct-write deposition of thermosensitive hydrogels is described as a facile route to obtain 3D cell-laden constructs with controlled 3D structure and stable behavior under physiological conditions.

Key words Thermogels, Direct-write deposition, 3D constructs, Tissue engineering (TE), Additive manufacturing (AM)

1 Introduction

Additive manufacturing (AM) of hydrogel-based materials has attracted growing interest as it enables the production of complex functional living tissues incorporating cells and/or bioactive molecules into three-dimensional structures. This emerging tool appears to be promising for advancing tissue engineering (TE) toward the fabrication of functional tissue and organ analogs for transplantation [1], as well as for drug screening and cancer or disease in vitro modeling [2, 3]. Thus, several research groups have adapted different AM techniques to generate cell-laden constructs [4, 5]. Such constructs can be obtained starting from a bioink, which is a suspension of cells in an aqueous solution based on hydrogels precursors, both of natural and synthetic origin, made insoluble in water through crosslinking processes (chemical or physical crosslinking). However, to achieve an accurate reproduction of the designed architecture, hydrogels have to meet specific requirements in terms of viscosity and gelation rate, which limits the number of formulations that can be processed by AM [6]. Moreover, the crosslinking process must be non-cytotoxic for embedded cells and should guarantee adequate structural integrity and mechanical

Alberto Rainer and Lorenzo Moroni (eds.), *Computer-Aided Tissue Engineering: Methods and Protocols*, Methods in Molecular Biology, vol. 2147, https://doi.org/10.1007/978-1-0716-0611-7_11,

properties for in vitro culture and in vivo implantation. Indeed, the low viscosity of the deposited fluid and the time necessary for its gelation often lead to low reproducibility, control, and resolution of the filament structure.

The authors have disclosed [7] the use of heat-sensitive hydrogels characterized by a sol-gel transition in a temperature range between 20 °C and 30 °C to prepare cell-laden constructs which are particularly stable under physiological conditions.

In the present work, the Pluronic/alginate gel system has been proposed as a prototypal thermosensitive gel to describe the procedure for the fabrication of cell-laden constructs with controlled 3D structure. This combination of polymers with distinct phase-transition mechanisms has been chosen with the aim of integrating the advantages of Pluronic gel in terms of printability [8] with the stability of the temperature-insensitive alginate component. The present system has been tested in combination with several cell lines (BALB/3 T3, C2C12, and Human Dermal Fibroblast) [9, 10] and successfully adopted for the engineering of muscle cell alignment through the use of C2C12 murine myoblast cell line [9].

2   Materials

To minimize the risk of contamination, AM equipment should be placed in a biological safety cabinet, and all the pieces of equipment in contact with the bioink should be properly sterilized before processing.

2.1 Preparation of the Bioink

1. Pluronic F127 (BASF).
2. Sodium alginate (PROTANAL® XP 3499, FMC BioPolymers).
3. Dulbecco's Modified Eagle Medium (DMEM).
4. Deionized water.
5. Autoclavable glass vial with lid and magnet.
6. Cell source (e.g., human dermal fibroblasts from adult skin, HDF) with appropriate medium and supplements.
7. Autoclave.
8. Ice bath.
9. Magnetic stirrer.

2.2 Additive Manufacturing

1. A three-axes motion control system (see **Note 1**).
2. Pressure-driven syringe barrel and piston system (Optimum system, Nordson EFD) terminated with a 250 μm ID blunt needle.

3. Heated holder for the pressure-driven syringe (operating at 37 °C) (*see* **Note 2**).

4. Programmable pressure controller (OB1-MK3, Elveflow, 0–8000 mbar range).

5. Petri dishes, glass slides, or other printing substrates.

6. Crosslinking solution: 25 mM $CaCl_2$

3 Methods

Although the procedures are described for a Pluronic/alginate solution of selected concentration, the proposed methods can be extended to each heat-sensitive hydrogel having a transition from the sol to the gel phase upon increasing the temperature, at values included between 20 °C and 30 °C. Information on the sol-gel transition kinetic, gelation time under physiological conditions should be determined for each novel printing ink. A detailed description of all these characterization steps is out of the scope of the present chapter. However, a representative example of the set of analyses required for the optimization of a novel thermogel can be found in Gioffredi et al. [8].

3.1 Preparation of Pluronic/Alginate Solution

The Pluronic/alginate solution can be prepared according to the so-called "cold method" proposed by Schmolka [11].

1. In a glass vial, add 100 mg of sodium alginate to 3.9 mL of an ice-cold 0.2× solution of DMEM under mild stirring.

2. While keeping the solution in an ice bath, slowly disperse 1 g of Pluronic F127 in the alginate solution under mild stirring (*see* **Note 3**).

3. Sterilize the solution by autoclaving, and store sterilized solution at 4 °C.

3.2 Printing Process

The main steps of scaffold manufacturing by direct-write deposition are summarized in Fig. 1.

Subconfluent HDFs can be routinely processed by trypsinization and centrifugation to obtain a cell pellet (*see* **Note 4**).

1. Working in a biosafety cabinet, homogeneously suspend HDFs in the sterile Pluronic/alginate solution kept at 4 °C, at a final concentration of 1×10^6 cells/mL.

2. Transfer the bioink into the pressure-driven syringe.

3. Mount the syringe on the heated holder, connect it to the pressure controller, and bring it at 37 °C (*see* **Note 5**).

4. Extrude the bioink at a pressure of 1.2 bar (*see* **Note 6**) with a needle/substrate relative speed of 10 mm/s. Several layers can

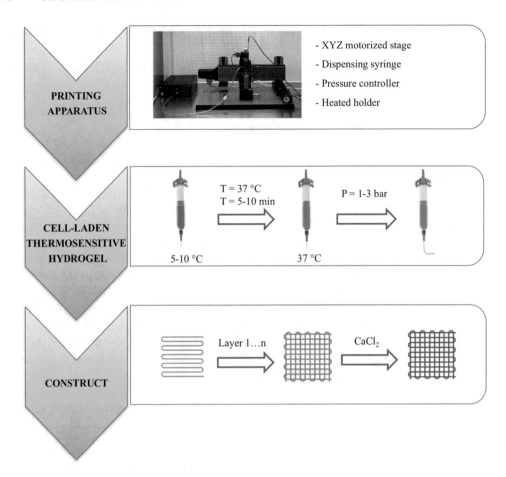

Fig. 1 Workflow for the fabrication of 3D constructs via direct-write deposition of thermosensitive hydrogel: (1) 3D printing apparatus, (2) syringe loading and sol-gel transition, (3) scaffold manufacturing and crosslinking

be superimposed to build a 3D construct of the desired geometry.

5. Gently dispense few mL of the crosslinking solution on the construct surface to induce alginate crosslinking.

6. After 5 min, remove the crosslinking solution, and gently wash the crosslinked scaffold in cell culture medium to terminate the gelation process.

7. Check the constructs under an inverted microscope. Two representative cell-laden constructs with different printing patterns are reported in Fig. 2.

8. Incubate constructs at 37 °C in a 5% CO_2 atmosphere with frequent medium changes (*see* **Note 7**). At selected timepoints, constructs can be retrieved and used for biological characterization.

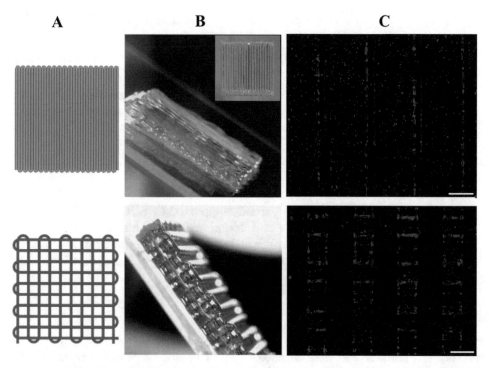

Fig. 2 Schematization of construct architectures (**a**), macrographs (**b**), and fluorescence micrographs (**c**) of the cell-laden constructs for two different geometries. HDF cells were labeled with PKH26 red-fluorescent dye. Scale bar: 500 μm

4 Notes

1. Several commercial or custom motion control systems can be adapted to direct-write deposition process. Please refer to [12] for an example of a custom-built apparatus.

2. This piece of equipment shall be customized upon the dimensions of the chosen syringe system to provide firm handling and homogeneous heating. Depending on the hydrogel composition, the use of an additional heated plate (set at 36–38 °C) as a holder for the printing substrate could be helpful to stabilize the gel structure after printing.

3. It is advisable to perform stepwise addition (5 or more aliquots) of Pluronic F127. Complete dissolution can take several hours.

4. The process can be adapted to different cell types, and the reported use of HDFs should be intended as a representative example. Given the mild conditions used in the processing method for preparing cell-laden constructs, the bioink can be supplemented with biomolecules of interests (e.g., growth factors).

5. The time necessary to induce bioink gelation is generally between 10 and 15 min.

6. Depending on the rheological properties of the printing solution, the extrusion pressure can be varied between 1 and 3 bar.

7. Pluronic F127 is rapidly eluted from the constructs during the first hours in culture and can be removed by frequent medium changes.

Acknowledgment

This work was partially supported by FIRB 2010 Project "Bioartificial materials and biomimetic scaffolds for a stem cells-based therapy for myocardial regeneration" (grant no. RBFR10L0GK) financed by MIUR (Italian Ministry of Education University and Research).

References

1. Ozbolat IT (2015) Bioprinting scale-up tissue and organ constructs for transplantation. Trends Biotechnol 33:395–400. https://doi.org/10.1016/j.tibtech.2015.04.005

2. Zhao Y, Yao R, Ouyang L et al (2014) Three-dimensional printing of Hela cells for cervical tumor model in vitro. Biofabrication 6:35001. https://doi.org/10.1088/1758-5082/6/3/035001

3. Vanderburgh J, Sterling JA, Guelcher SA (2017) 3D printing of tissue engineered constructs for in vitro modeling of disease progression and drug screening. Ann Biomed Eng 45:164–179. https://doi.org/10.1007/s10439-016-1640-4

4. Yoo S-S (2009) 3-dimensional multi-layered hydrogels and methods of making the same, US Patent 13/063,502

5. Hasan SK (2009) Three dimensional tissue generation, US Patent 13/130,740

6. Melchels FPW, Klein TJ, Malda J et al (2012) Additive manufacturing of tissues and organs. Prog Polym Sci 37:1079–1104. https://doi.org/10.1016/J.PROGPOLYMSCI.2011.11.007

7. Chiono V, Sartori S, Boffito M et al (2016) Method for preparing cellularized constructs based on heat-sensitive hydro-gels. Italian Patent IT102015000020718

8. Gioffredi E, Boffito M, Calzone S et al (2016) Pluronic F127 hydrogel characterization and biofabrication in cellularized constructs for tissue engineering applications. Procedia CIRP 49:125–132. https://doi.org/10.1016/j.procir.2015.11.001

9. Mozetic P, Giannitelli SM, Gori M et al (2017) Engineering muscle cell alignment through 3D bioprinting. J Biomed Mater Res A 105:2582–2588. https://doi.org/10.1002/jbm.a.36117

10. Giannitelli SM, Mozetic P, Trombetta M et al (2015) Additive manufacturing of pluronic/alginate composite thermogels for drug and cell delivery. In: Srivatsan TS, Sudarshan TS (eds) Additive manufacturing: innovations, advances and applications. CRC Press, Boca Raton

11. Schmolka IR (1977) A review of block polymer surfactants. J Am Oil Chem Soc 54:110–116. https://doi.org/10.1007/BF02894385

12. Rainer A, Giannitelli SM, Accoto D et al (2012) Load-adaptive scaffold architecturing: a bioinspired approach to the design of porous additively manufactured scaffolds with optimized mechanical properties. Ann Biomed Eng 40:966–975. https://doi.org/10.1007/s10439-011-0465-4

Chapter 12

3D Bioprinting of Complex, Cell-laden Alginate Constructs

Atabak Ghanizadeh Tabriz, Dirk-Jan Cornelissen, and Wenmiao Shu

Abstract

Biofabrication has been receiving a great deal of attention in tissue engineering and regenerative medicine either by manual or automated processes. Different automated biofabrication techniques have been used to produce cell-laden alginate hydrogel structures, especially bioprinting approaches. These approaches have been limited to 2D or simple 3D structures, however. In this chapter, a novel bioprinting technique is disclosed for the production of more complex alginate hydrogel structures. This was achieved by dividing the alginate hydrogel cross-linking process into three stages: primary calcium ion cross-linking for printability of the gel, secondary calcium ion cross-linking for rigidity of the alginate hydrogel immediately after printing, and tertiary barium ion cross-linking for the long-term stability of the alginate hydrogel in the culture medium.

Key words 3D bioprinting, Alginate hydrogel, Biofabrication, Bioextrusion

1 Introduction

Three-dimensional (3D) bioprinting has drawn great attention in tissue engineering as an emerging technology during the past decade for the regeneration and restoration of damaged or lost tissues and organs [1–3] such as bones, skin, nose, and human ear [4–7].

Several bioextrusion [8–10] printing methods have been developed to print living cells into a tissue or organ-like structure; however, these structures are limited in complexity and scale.

The main challenge is balancing the printing conditions in such a manner that bioscaffold material is viscous enough to generate a rigid 3D structure while minimizing damage to the cells.

In this chapter, a new bioprinting technique using a modified Fab@Home 3D printer for complex 3D alginate hydrogel structures is presented [11] as shown in Fig. 1. The technique uses a three-step cross-linking process to ensure the printability, rigidity, and stability of the gel that can be gentle to cells during the bioprinting process. Alginate hydrogels were chosen because of

Alberto Rainer and Lorenzo Moroni (eds.), *Computer-Aided Tissue Engineering: Methods and Protocols*,
Methods in Molecular Biology, vol. 2147, https://doi.org/10.1007/978-1-0716-0611-7_12,
© Springer Science+Business Media, LLC, part of Springer Nature 2021

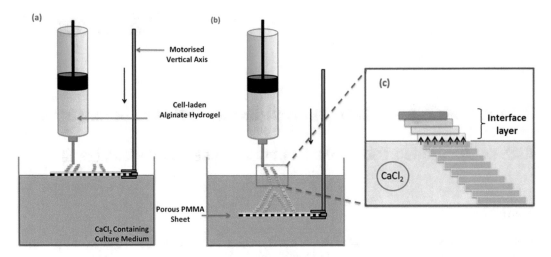

Fig. 1 Schematic drawing of printing partially cross-linked alginate hydrogel. (**a**) Initially, few layers of partially cross-linked alginate are printed on top of the porous membrane, until the limit height for structural integrity is reached. (**b**) Further cross-linking is needed once the structure is about to collapse by lowering the Z stage into the secondary crosslinking bath to create a support for the upcoming layers. (**c**) The interface layers: upward diffusion of Ca^{+2} ions into the interface layers which are partially cross-linked above the $CaCl_2$ solution [11]

their biocompatibility and ability to form highly porous structures with the ability to support cell proliferation [2, 12, 13].

2   Materials

2.1   Alginate and Cross-linking Reagents

1. Sodium alginate solution: 8% (w/v) sodium alginate in deionized water (DIW) (*see* **Note 1**).

2. Primary crosslinking solution: 80 mM calcium chloride in DIW; for printing cell-laden constructs, concentration shall be increased to 160 mM.

3. Secondary crosslinking solution: 100 mM calcium chloride in DIW.

4. Tertiary crosslinking solution: 40 mM barium chloride in DIW.

2.2   Equipment and Consumables for 3D Printing

1. 5 mL syringes.

2. Dispensing needle.

3. Nitrocellulose filtration paper.

4. Petri dishes.

5. Tweezers.

6. Vortex.

7. 3D bioprinter (e.g., Fab@home) (*see* **Note 2**).

8. Computer with CAD and 3D printing softwares.

9. Cell suspension: ca. 20 million cells per mL (*see* **Notes 3** and **4**).

10. Cell culture hood.

11. Incubator.

12. Heated bath.

13. Gamma irradiator.

14. Autoclave.

15. Live/dead assay: 6 mg/mL fluorescein stock solution to stain live cells and 2.5 mM propidium iodide stock solution to stain dead cells (*see* **Note 5**).

3 Methods

3.1 Alginate Hydrogel and Cross-Linking Reagents Sterilization for 3D Printing

1. Sterilize the alginate solution by gamma irradiation with 10 Gy at a rate of 1 Gy/min.

2. Sterilize the cross-linking reagents by autoclaving them at 121 °C for 15 min.

3.2 Design and 3D Printing of Alginate Hydrogel Structures

1. Prepare the desired 3D CAD and convert it to STL format for printing.

2. Open the STL file through the relative 3D printer software (*see* **Notes 6** and **7**).

3. Set printing width to 0.35 mm, the printing height to 0.475 mm, and the printing speed to 6 mm/s.

4. Generate the 3D printing file (XDFL format, for the present setup).

3.3 3D Printing Procedure

1. Add 80 mM $CaCl_2$ solution to the alginate solution in a centrifuge tube at a 1:1 volume ratio and shake for 10 s to form a partially cross-linked alginate hydrogel.

2. Further vortex the gel at 1500 rpm for 30 s.

3. Load some alginate into the extrusion syringe using a 5 mL syringe.

4. Dampen the nitrocellulose filter paper with the secondary crosslinking solution and place it into a sterile Petri dish.

5. Purge some gel to ensure the continuous printing of the gel.

6. Open the 3D printing software and execute printing.

7. Once two layers of the 3D structure have been printed, gently add some secondary crosslinking solution (100 mM $CaCl_2$) into the Petri dish until the level reaches the uppermost surface of the printed structure. Make sure the secondary crosslinking bath does not overflow the printed structure (*see* **Note 9**).

8. Repeat **step 5** every two layers until completion of the printing sequence.

9. Once the 3D printing is completed, leave the printed structure in the secondary crosslinking bath for 10 min.

10. Gently remove the secondary crosslinking solution with a syringe and replace it with the tertiary crosslinking solution.

11. After 2 min, gently remove the printed structure from the tertiary crosslinking bath using tweezers.

3.4 Bioprinting of Cell-laden Alginate Hydrogel

1. Clean the 3D bioprinter using 70% ethanol and place it inside a cell culture hood.

2. Heat up the sodium alginate solution and both the primary and secondary crosslinking solutions to 37 °C in a heated bath.

3. Inside of a cell culture hood, load 2 mL of sterile sodium alginate solution into the extrusion syringe.

4. Load 1 mL of sterile primary crosslinking solution (160 mM $CaCl_2$) into the extrusion syringe.

5. Load 1 mL of cell suspension into the extrusion syringe.

6. Gently mix the three solutions using a pipet tip until the gel is formed (2–3 min).

7. Load the syringe in the bioprinter, and continue the printing process as described in Subheading 3.3, ensuring that you work in sterility.

8. After printing, transfer the structure to a Petri dish containing cell culture medium, and place in an incubator at 37 °C and 5% CO_2. Refresh medium every 2 days.

3.5 Live and Dead Cell Assay

After printing the cells, it is important to keep an eye out for cell survival. The following protocol was used for routine cell viability checks (*see* **Note 8**).

1. In a cell culture hood, aspirate the medium from the construct and replace with PBS.

2. Add 1 μL of propidium iodide stock solution for every mL of PBS and incubate in the dark at 37 °C for 30 min.

3. After incubation, add 0.5 μL fluorescein diacetate stock solution per mL of PBS, and incubate at room temperature for 3–5 min.

4. Aspirate the staining solution and wash the structure once with PBS.

5. Image the constructs using a fluorescent microscope. Live cells will be green fluorescent (FITC filter set), dead cells will have red fluorescent nuclei (TRITC filter set) (*see* **Note 7**).

4   Notes

1. Alginate solution can be prepared in several ways. If using a centrifuge tube, use the ultrasonic bath to fasten the process. If using a glass beaker, make sure the glass beaker is covered and put into a mixer overnight at 180 rpm and at room temperature.

2. The present procedure has been optimized for Fab@Home (https://reprap.org/wiki/RUG/Pennsylvania/State_Col lege/Printers/PSU_Fab@Home) printing hardware and Seraph studio (https://github.com/SeraphRobotics) printing software. However, the procedure can easily be adapted to different printing setups.

3. In the present experiments, we have used a sturdy carcinoma cell lines. However, biofabrication using alginate can also be used with more sensitive cells. To work with more sensitive cells, we recommend the following modifications:

   (a) Lower the percentage of alginate. This will result in less mechanical strength of the structure, which means it will not be possible to print structures that are more than a few millimeters high. When lowering the percentage of alginate, also lower the concentration of the primary $CaCl_2$ crosslinking solution. When working with a 4% w/v alginate solution, use a 40 mM primary crosslinking solution.

   (b) Use water for embryo transfer for the preparation of the alginate solution. As lab-grade DIW can get quite acidic, we use water for embryo transfer with our more delicate cells. This increases cell survival after printing. The water can also be used for the preparation of the crosslinking solutions.

   (c) When exploring different cell concentration levels, take into account that the cells will have an influence on the stability of the structure, especially for very high cell concentrations.

4. Take care that the medium used to create the cell suspension in Subheading 3.4 does not contain any large quantity of cations that might react with the alginate.

5. This staining protocol has been optimized for application with our alginate structures. When using commercial live/dead assays containing ethidium homodimer-1 and calcein, we found that our structures were less stable, or even dissolved, after staining.

6. When designing structures make sure the thickness for the structures is at least twice the printing path to avoid errors while generating the printing sequence. The minimum thickness the structure will have after printing is roughly 1 mm due to swelling of alginate after cross-linking.

7. Printing parameters can be varied using different diameter printing tips. The mentioned printing parameters are for a 0.35 mm diameter tip.

8. When it is not necessary to do a life and dead assay on the entire construct, use sterile scalpels to cut off a small piece of the construct, and transfer it into a new plate using sterile tweezers before the staining protocol.

9. Alternatively, the nitrocellulose membrane can be placed on a motorized z-stage, that can be programmatically lowered to submerge the construct in the secondary crosslinking solution while it is additively manufactured, as depicted in Fig. 1.

Acknowledgments

AGT acknowledges the scholarship support by Heriot-Watt University. DJC acknowledges the scholarship from Medical Research Scotland.

References

1. Mironov V, Trusk T, Kasyanov V, Little S, Swaja R, Markwald R (2009) Biofabrication: a 21st century manufacturing paradigm. Biofabrication 1(2):022001

2. Wang C, Tang Z, Zhao Y, Yao R, Li L, Sun W (2014) Three-dimensional in vitro cancer models: a short review. Biofabrication 6(2): 022001

3. Huang G, Wang L, Wang S, Han Y, Wu J, Zhang Q, Xu F, Lu TJ (2012) Engineering three-dimensional cell mechanical microenvironment with hydrogels. Biofabrication 4 (4):042001

4. Reiffel AJ, Kafka C, Hernandez KA, Popa S, Perez JL, Zhou S, Pramanik S, Brown BN, Ryu WS, Bonassar LJ, Spector JA (2013) High-fidelity tissue engineering of patient-specific auricles for reconstruction of pediatric microtia and other auricular deformities. PLoS One 8 (2):e56506

5. Boland T, Tao X, Damon BJ, Manley B, Kesari P, Jalota S, Bhaduri S (2007) Drop-on-demand printing of cells and materials for designer tissue constructs. Mater Sci Eng C 27(3):372–376

6. Hollister SJ (2005) Porous scaffold design for tissue engineering. Nat Mater 4(7):518–524

7. Koch L, Deiwick A, Schlie S, Michael S, Gruene M, Coger V, Zychlinski D, Schambach A, Reimers K, Vogt PM, Chichkov B (2012) Skin tissue generation by laser cell printing. Biotechnol Bioeng 109 (7):1855–1863

8. Cui X, Boland T (2009) Human microvasculature fabrication using thermal inkjet printing technology. Biomaterials 30(31):6221–6227

9. Shim J-H, Kim JY, Park M, Park J, Cho D-W (2011) Development of a hybrid scaffold with synthetic biomaterials and hydrogel using solid freeform fabrication technology. Biofabrication 3(3):034102

10. Shim J-H, Lee J-S, Kim JY, Cho D-W (2012) Bioprinting of a mechanically enhanced three-dimensional dual cell-laden construct for osteochondral tissue engineering using a multi-head tissue/organ building system. J Micromech Microeng 22(8):085014

11. Tabriz G, Hermida MA, Leslie N, Shu W (2015) Three-dimensional bioprinting of complex cell laden alginate hydrogel structures. Biofabrication 7(4):045012

12. Zhao Y, Yao R, Ouyang L, Ding H, Zhang T, Zhang K, Cheng S, Sun W (2014) Three-dimensional printing of Hela cells for cervical tumor model in vitro. Biofabrication 6:035001

13. Liliang O, Rui Y, Shuangshuang M, Xi C, Jie N, Wei S (2015) Three-dimensional bioprinting of embryonic stem cells directs highly uniform embryoid body formation. Biofabrication 7:044101

Chapter 13

Surface Tension-Assisted Additive Manufacturing of Tubular, Multicomponent Biomaterials

Elia A. Guzzi, Héloïse Ragelle, and Mark W. Tibbitt

Abstract

The fabrication of functional biomaterials for organ replacement and tissue repair remains a major goal of biomedical engineering. Advances in additive manufacturing (AM) technologies and computer-aided design (CAD) are advancing the tools available for the production of these devices. Ideally, these constructs should be matched to the geometry and mechanical properties of the tissue at the needed implant site. To generate geometrically defined and structurally supported multicomponent and cell-laden biomaterials, we have developed a method to integrate hydrogels with 3D-printed lattice scaffolds leveraging surface tension-assisted AM.

Key words Multicomponent biomaterials, Additive manufacturing, Computer-aided design, Regenerative medicine, 3D printing

1 Introduction

Living biomaterials for organ replacement or tissue repair are fabricated traditionally by placing cells on or within engineered support matrices [1, 2]. Ideally, these constructs should be designed to match the geometry and mechanical properties of the native tissue at the implant site in the individual patient and provide a suitable microenvironment for cell growth and tissue integration. The design and fabrication of such personalized implants is being enabled by advances in computer-aided design (CAD) and additive manufacturing (AM) technologies combined with new chemistries for the fabrication of materials amenable to biomedical applications [3].

3D bioprinting is an emerging tool for manufacturing living biomaterials for regenerative medicine that allows for tailored geometry and cellular organization specific to the tissue of interest [4]. 3D bioprinting leverages techniques developed for AM to fabricate, layer-by-layer, spatially organized and complex

Alberto Rainer and Lorenzo Moroni (eds.), *Computer-Aided Tissue Engineering: Methods and Protocols*,
Methods in Molecular Biology, vol. 2147, https://doi.org/10.1007/978-1-0716-0611-7_13,
© Springer Science+Business Media, LLC, part of Springer Nature 2021

biomaterials made of hydrogels and cells. Common strategies for 3D bioprinting include inkjet bioprinting, microextrusion, laser-assisted bioprinting, and freeform reversible embedding of suspended hydrogels [3–5]. In the 3D bioprinting paradigm, CAD is used to design the final construct and the associated printing technology is used to render it. One limitation of these techniques is that hydrogel-based biomaterials can demonstrate insufficient mechanical stability or structural integrity for some applications as they are often printed with relatively low modulus materials ($E < 100$ kPa).

A significant advance in 3D bioprinting for tissue replacements was the development of multimaterial printers, such as the integrated tissue-organ printer (ITOP) [6]. This technology enabled the fabrication of cell-laden tissue constructs with sufficient mechanical integrity through sequential printing of cell-laden hydrogels along with structural support polymers, e.g., polycaprolactone (PCL). The incorporation of PCL within the printing process allowed the ITOP to fabricate stable and cell-laden tissue constructs on the human-scale. While transformative, multimaterial printers require complex and specialized instrumentation making the costs prohibitive for some labs, start-up companies, and broad clinical use. In addition, cell viability can be impacted by long-cell handling times and the whole fabrication process must be conducted within a controlled and sterile environment. Therefore, new technologies are needed to extend these advances in multicomponent biomaterials and computational design in order to accelerate the translation of personalized medical implants.

To complement existing approaches, we have developed a facile method to engineer multicomponent biomaterials with predefined geometry and mechanical properties [7]. The general approach exploits CAD and traditional AM to design and fabricate a fenestrated lattice with defined geometry and mechanical properties. Then, surface tension is used to coat the 3D-printed lattice with suspended liquid films that can be transformed subsequently into solid hydrogels, which can optionally contain encapsulated cells. This multi-stage approach enables the rapid production of multimaterial and cell-laden tissue constructs on the mm to cm lengthscales and constitutes an additional unit operation in AM beyond layer-by-layer deposition. The 3D-printed lattice controls the geometry and physical properties of the final biomaterial while the hydrogel offers a substrate for cell seeding, encapsulation, and biointegration. In addition, surface tension-assisted AM avoids some challenges with traditional 3D bioprinting by enabling minimal cell handling and decreased time for fabrication as material does not need to be deposited at each voxel in the final design. In total, this method based on surface tension-assisted AM comprises a facile approach to design and fabricate multicomponent biomaterials with predefined geometry, tailored mechanical properties, and tunable biological function.

The surface tension-assisted AM method is particularly useful for the generation of hollow structures. Here, we describe the method for the fabrication of tubular, multicomponent biomaterial constructs. We detail the design and fabrication of the lattices via traditional AM and the coating of the constructs with a methacryloyl gelatin hydrogel that can contain mammalian cells. Two 3D-printed lattices are described that provide the final device with different mechanical properties. Finally, we discuss methods to characterize the device after fabrication. While we focus on one specific use of the technology here, the method is versatile and can be used for the fabrication of other geometries, with different 3D printing technologies, materials, and applications as discussed in many of the notes (*see* Subheading 4).

2 Materials

This method assumes access to normal wet lab facilities and materials such as pipettes, pipette tips, standard glassware, tweezers, and spatulas.

2.1 General Lab Equipment

1. Sterile 15 and 50 mL centrifuge tubes.
2. Heat block for 15 and 50 mL centrifuge tubes.
3. 100-mm polystyrene Petri dishes (*see* **Note 1**).
4. Hot plate or magnetic stirrer with heat.
5. Biosafety cabinet and CO_2 incubator.
6. Universal mechanical testing machine.

2.2 3D Printing

1. 3D printer: VIPER si2™ SLA® SYSTEM (3D Systems) (*see* **Note 2**).
2. AutoCAD® software (Autodesk); MeshMixer (Autodesk) (*see* **Note 3**).
3. Resin: Accura® ClearVue™ (3D Systems) (*see* **Note 4**).

2.3 Hydrogel Solution

1. Gelatin methacryloyl bloom 300, 80% degree of substitution (GelMA; Sigma-Aldrich) (*see* **Note 5**).
2. Lithium phenyl-2,4,6-trimethylbenzoylphosphinate (LAP) (*see* **Note 6**).
3. Distilled water—molecular biology grade.
4. GelMA stock solution: Prepare a sterile stock solution of GelMA at 10% (wt/wt) in distilled water. In order to dissolve GelMA, the solution needs to be heated to 37 °C in a heat block. Store GelMA stock solution at 4 °C.

5. LAP stock solution: Prepare a sterile stock solution of LAP at 5% (wt/wt) in distilled water. Store aliquots of LAP stock solution at −20 °C for short-term storage and at −80 °C for long-term storage.

6. Hydrogel precursor solution: Prior to coating the 3D-printed lattice, prepare the hydrogel precursor solution from the GelMA and LAP stock solutions at a final concentration of 7.5% (wt/wt) GelMA and 0.5% (wt/wt) LAP. The balance of volume can be comprised of distilled water or concentrated cell solutions, in the case of direct cell encapsulation. The hydrogel precursor solution should be prepared immediately prior to use and maintained at 37 °C as the solution will gel upon cooling (*see* **Note 7**).

2.4 Coating and Photopolymerization

1. Hydrogel precursor solution: Prepared fresh as described in Subheading 2.3, **item 6**. The hydrogel precursor solution should be used within 2 h of preparation.

2. Collimated, LED UV light source ($\lambda = 365$ nm, $I_0 \sim 6.0$ mW/cm^2) (*see* **Note 8**).

3. UV blocking safety glasses.

4. Opaque enclosure for photopolymerization to protect the user and environment from UV light.

2.5 Cell Culture

1. MRC-5 lung fibroblast cells (ATCC® CCL-171) and human umbilical vein endothelial cells (HUVEC; ATCC® PCS-100-013).

2. Cell culture medium for MRC-5: Minimal Essential Medium α (αMEM) supplemented with 10% heat-inactivated fetal bovine serum (HI-FBS) and 1% Penicillin-Streptomycin (Pen-Strep; 10,000 U/mL).

3. Collagen I pre-coated tissue culture plates (*see* **Note 9**).

4. Cell culture medium for HUVEC: Endothelial Cell Growth Medium-2 (EGM2; Lonza) supplemented with 10% HI-FBS and 1% Pen-Strep. HUVECs must be grown on collagen I-coated tissue culture plates.

5. Cell culture medium for co-culture: EGM2 supplemented with 10% HI-FBS and 1% Pen-Strep.

6. Collagen solution: collagen I, rat tail (50 µg/mL) in 0.02 N acetic acid in distilled water.

7. LIVE/DEAD® Viability/Cytotoxicity Kit, for mammalian cells (ThermoFisher).

2.6 Confocal Microscopy

1. Confocal microscope equipped with 10× air and 30× oil immersion objectives.

2. 32% Paraformaldehyde (formaldehyde) aqueous solution.

3. NucBlue™ Fixed Cell ReadyProbes™ Reagent (ThermoFisher).

4. AlexaFluor® 488 Phalloidin (AF488-Phalloidin; ThermoFisher).

5. CellTracker™ Red CMTPX Dye (ThermoFisher).

6. Fixative solution: 3.2% Paraformaldehyde solution prepared by diluting 32% Paraformaldehyde aqueous solution 1:9 with Hank's Buffered Saline Solution supplemented with calcium and magnesium (HBSS +/+).

7. Staining solution: Prior to imaging, prepare a fresh stock solution with NucBlue (2 drops per mL) and AF488-Phalloidin (5 µg/mL) in HBSS +/+.

3 Methods

3.1 Scaffold Design

Different options are currently available for generating a suitable 3D model for AM in biomedical engineering. On one hand, the user can design a construct using CAD software and principle features of the target tissue. The advantages of this method are the freedom of designing geometries with specific features and the ability to break down, or combine, features into multiple components. On the other hand, the user can combine high-resolution medical images, such as computed tomography (CT), micro-CT, or magnetic resonance imaging (MRI), with CAD to design a construct informed by anatomical information about the native tissue at the implant site. The generation of a 3D model from medical images requires post-processing and usually starts with image segmentation [8]. In this step, the volumetric image data can be filtered and partitioned into multiple segments, or pixels, which define the boundaries and spatial arrangement of objects in the image. Then, the volume representation of the 3D image can be applied for volume rendering and generation of anatomic models [9]. Moreover, the segmented objects can be used to create a 3D triangle mesh, in which triangular facets approximate the external shape of the object (*see* **Note 10**). One of the main advantages of having a 3D image containing labeled objects is the ease of engineering a 3D model with specific anisotropy and mechanical properties that can be useful to predict its mechanical deformation using finite element (FE) simulation.

In this description of the method, we employ CAD software to design tubular constructs without medical imaging:

1. Draw the tubular scaffolds in AutoCAD® software (Autodesk, San Rafael, CA, USA) with internal diameter 10 mm and external diameter 12 mm and longitudinal length 30 mm with two different window geometries (Fig. 1):

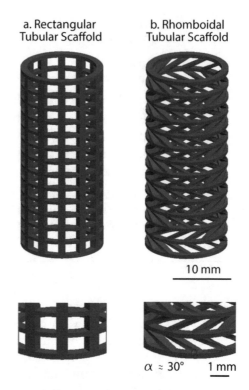

Fig. 1 Geometry of the 3D-printed and fenestrated tubular scaffolds (height = 30 mm, external diameter = 12 mm, internal diameter = 10 mm). Each layer ($n = 13$) is designed with 12 windows along the circumference. The first and last ring are 0.625 mm thick, whereas all other pipes are 0.5 mm thick. The vertical wall between each window is 1.04 mm. (**a**) The rectangular windows are 2.09 mm wide (dimension of the arc on the external surface of the cylinder) and 1.75 mm high. (**b**) The rhomboidal windows were designed with the same dimension of the rectangular windows ($w = 2.09$ mm, $h = 1.75$ mm), and additionally tilted with an angle α of approximately 30°

(a) Rectangular shaped windows with external arc length 2.09 mm and height 1.75 mm composed of 12 windows in each circumferential layer and 13 windows in each longitudinal column.

(b) Rhomboidal shaped windows with external arc length 2.09 mm, height 1.75 mm, and angle α ~30° composed of 12 windows in each circumferential layer and 13 windows in each longitudinal column.

2. Export the generated 3D models as stereolithography files (Surface Tesselation Language, STL).

3. Post-processing of the digital model with MeshMixer might be necessary in order to refine the mesh, such as repairing discontinuities, smoothing staircasing errors, or even to combine or remove multiple structures (*see* **Note 11**).

3.2 Scaffold Fabrication

Multiple 3D printing technologies are available for the scaffold fabrication, each offering specific advantages and disadvantages. For this method, stereolithography (SLA; Accura® ClearVue™; VIPER si2™ SLA®, 3D Systems) was used, but the workflow can be adapted to other systems, such as fused deposition modeling (FDM) (*see* **Note 12**).

1. Open the 3D model in the printer software.

2. Define the printing orientation of the model in order to get the best quality and to reduce the support material required for overhanging geometry. In this procedure, the tubular structures were oriented longitudinally.

3. Based on the laser diameter and resin solution define the printing parameters, such as layer thickness, hatch spacing, fill spacing, infill pattern, and density. To produce the tubular scaffolds, the fenestrated mesh-like structures were printed with Accura® ClearVue™. The printed layers height was set to 0.004″ (ca. 100 μm) as the laser diameter used was 100 μm.

4. Once all of the required parameters have been defined, execute the slicing of the model and export the G-code file containing the printing instructions.

5. Load the file on the VIPER si2™ SLA® 3D printer and initiate the printing.

6. At the end of the process, clean the 3D-printed construct with isopropyl alcohol to remove the excess resin, dry and post-process with UV light ($\lambda = 365$ nm, $t = 30$ min).

7. In case the scaffolds are printed with support material, remove it after UV-curing. Moreover, to smooth the lattice surface and eliminate support marks, the printed object can be rinsed with water and/or water mixed with sand.

3.3 Hydrogel Preparation

1. If encapsulating MRC-5 lung fibroblasts in the hydrogel, prepare a concentrated cell suspension of MCR-5 lung fibroblasts in the culture medium. The concentration of the suspension will depend on the confluency and number of plates available.

2. Prepare a fresh batch of hydrogel precursor solution under sterile conditions (Subheading 2.3). If encapsulating MRC-5 lung fibroblasts, add the appropriate volume of the concentrated cell suspension to achieve a final concentration of MRC-5 lung fibroblasts of 1.0×10^6 cells/mL. Triturate to ensure homogeneity of the cells in the hydrogel precursor solution.

3. Maintain the hydrogel precursor solution at 37 °C during use. The prepared solution should be used with ~2 h without cells and as quickly as possible if cells are suspended in the precursor solution.

3.4 Hydrogel Coating

1. Autoclave the scaffold prior to coating to ensure sterility. All steps of the hydrogel coating should be performed under sterile conditions, such as in a biosafety cabinet.

2. Place the lid of a 100-mm polystyrene Petri dish on a hot plate at 37 °C.

3. Pipette 1 mL of the hydrogel precursor solution onto the lid of the 100-mm Petri dish (Fig. **2**).

4. Place the fenestrated tubular lattice into the solution horizontally.

5. The windows in contact with the fluid should coat with suspended liquid films immediately. With a pair of tweezers, gently rotate the tubular construct 1–2 full rotations through the solution, maintaining the lattice in a horizontal position.

6. Visually inspect that all of the windows have been coated successfully. If not, rotate the device through the solution again.

7. Withdraw the coated lattice carefully from the precursor solution and expose the coated lattice to UV light ($\lambda = 365$ nm; $I_o = 6.0$ mW/cm^2; $t = 120$ s) to crosslink the suspended liquid films.

8. Upon gelation, place the tubular, multicomponent construct in HBSS +/+ or culture medium.

3.5 Secondary Seeding of Cells

1. Label a plate of HUVECs with CellTracker™ Red CMTPX Dye according to the manufacturer's protocol.

2. Prepare a cell suspension of HUVECs at a concentration of 1.5×10^6 cells/mL in the culture medium.

3. Inside a biosafety cabinet, remove the hydrogel-coated tubular scaffold from HBSS +/+ or culture medium and plug one end with a 3D-printed plug (10 mm diameter) by sealing it to the base of the tubular construct with cell-free hydrogel precursor solution.

4. Stand the tubular construct on end longitudinally with the sealed end down and fill with ~1 mL collagen solution and let stand for 20 min prior to cell seeding.

5. Gently aspirate the collagen solution and rinse with ~1 mL of HBSS +/+. Gently aspirate the HBSS +/+.

6. With the plug still in place, rest the tubular construct on its side horizontally and add ~100 μL of the HUVEC suspension (1.5×10^6 cells/mL). Let sit for ~10–15 min. Rotate the device 90° and add another ~100 μL of the HUVEC suspension and let sit for ~10–15 min. Repeat until the whole lumen of the construct has been coated with HUVECs.

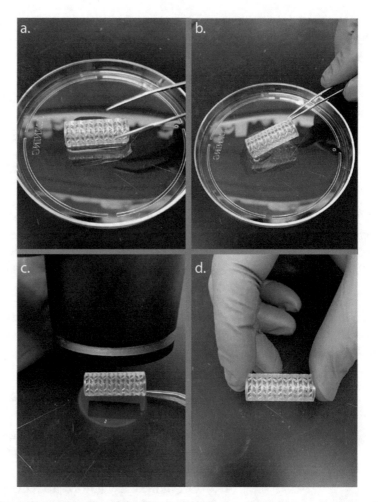

Fig. 2 Hydrogel coating of the 3D-printed lattice scaffolds. (**a**) A small volume of hydrogel precursor solution is placed on the lid of a 100-mm Petri dish and the scaffold is placed in the solution horizontally to allow surface tension to suspend liquid films across the windows of the lattice support. (**b**) The scaffold is rotated through in the solution to ensure even coating of all of the windows and then gently removed from the hydrogel precursor solution. (**c**) The suspended liquid films are transformed into a solid hydrogel coating by exposing the device to UV light ($\lambda = 365$ nm; $I_0 = 6.0$ mW/cm^2; $t = 120$ s). (**d**) The resultant tubular, multicomponent biomaterial is produced after photopolymerization combining the 3D-printed support with solid hydrogel films in each of the windows

7. Gently place in the culture medium. The plug can be removed gently after 24 h.

8. To maintain the co-culture device, change the culture medium every 2–3 days or as needed.

3.6 Mechanical Characterization of the Constructs

To assess the mechanical properties of the formed constructs, perform radial and longitudinal uniaxial compression tests for devices with both rectangular or rhomboidal windows.

1. Set the displacement rate to 3.0 mm/min and the total displacement to 6.0 mm and 10.0 mm for radial and longitudinal tests, respectively.

2. Calculate the effective modulus under uniaxial compression from the slope of the linear fit on the first 0.5 mm of the compression stress-strain curve and the known geometry of the devices.

3.7 Biological Characterization of the Constructs

1. Monitor cell viability with LIVE/DEAD® Viability/Cytotoxicity kit at desired timepoints after encapsulation. Calculate the viability percentages as the ratio between the number of viable cells (labeled with Calcien AM; green) and the total number of cells in the sample (*see* **Note 13**).

3.8 Microscopic Characterization of the Constructs

Use confocal microscopy to characterize cell spreading within the hydrogel windows and coating on the construct surface.

1. Remove the construct from the culture medium and rinse with HBSS +/+.

2. Fix the samples with 3.2% Paraformaldehyde for 15 min at room temperature.

3. Immerse the scaffold in the staining solution for 2 h at 4 °C to stain the cell nuclei with NucBlue and the actin filaments with AF-488 Phalloidin.

4. Isolate individual windows from the constructs for imaging and immerse each in a glass-bottomed imaging chamber containing HBSS +/+.

5. Collect confocal Z-stacks with a 10× air or a 30× oil immersion objective to characterize the full thickness of the window.

6. Reconstruct the Z-stacks in ImageJ to generate 2D projections or 3D renderings.

4   Notes

1. The desired diameter of the Petri dish will depend on the size of 3D-printed lattice to be coated. For the devices described here, 100-mm Petri dishes were sufficiently large.

2. Various SLA printers are currently available on the market and can be used as alternatives to the VIPER si2™ SLA® SYSTEM (3D Systems).

3. Other CAD software (e.g., Pro/Engineering, Siemens NX, SolidWorks) as well as mesh modifying and slicing software (e.g., MeshLab, Slic3r, Simplify3D) can be used.

4. The choice of the resin used in SLA can vary depending on the need of the mechanical properties aimed, required compatibility with biological systems, and SLA printer used.

5. The degree of methacrylation on the gelatin can be modified to alter the final mechanical properties of the hydrogel [10]. Moreover, other liquid precursor solutions can be used as coating material, such as alginate, collagen or collagen/elastin, and neat thiol-ene solutions.

6. LAP can also be synthesized following the literature protocol [11]. Other photoinitiators, such as 1-[4-(2-hydroxyethoxy)-phenyl]-2-hydroxy-2-methyl-1-propanone (Irgacure 2959), can be used to trigger the photopolymerization reaction in combination with UV light.

7. The initial concentration of the hydrogel precursor solution can be modified to alter the final mechanical properties of the hydrogel.

8. Other options for UV light sources are available and depend on the selected photoinitiator as well as the specific wavelengths needed to initiate the photopolymerization reaction.

9. Collagen I-coated plates can be prepared by placing a 50 μg/mL solution of collagen I, rat tail in 0.02 N acetic acid in distilled water on standard tissue culture polystyrene plates for 1 h at room temperature (e.g., ~7 mL for a standard T75 flask). Prior to use for cell culture, rinse with PBS.

10. Software available for segmenting digital images includes Seg3D, 3D Slicer, ITK-SNAP, ImageJ, and InVesalius. Other software such as MedCAD, Reverse engineering interface, and STL interface can be used to convert medical images directly to CAD models [8, 9].

11. Other software can be used for this purpose, see **Note 3**. Moreover, other geometries can be fabricated, such as planar mesh-like scaffolds on the mm to cm length-scales [7].

12. The current workflow can be adapted to other systems, such as fused deposition modeling (FDM), melt electrospinning, and selective laser melting (SLM). Based on the fabrication system applied, different resolutions can be achieved.

13. Following treatment with the LIVE/DEAD® kit, imaging should be conducted within ~1 h.

Acknowledgment

This work was supported by startup funds from ETH Zürich.

References

1. Langer R, Vacanti JP (1993) Tissue engineering. Science 260(5110):920–926
2. Tibbitt MW, Langer R (2017) Living biomaterials. Acc Chem Res 50(3):508–513
3. Zhang YS, Oklu R, Dokmeci MR, Khademhosseini A (2018) Three-dimensional bioprinting strategies for tissue engineering. Cold Spring Harb Perspect Med 8(2). https://doi.org/10.1101/cshperspect.a025718
4. Murphy SV, Atala A (2014) 3D bioprinting of tissues and organs. Nat Biotechnol 32:773–785
5. Hinton TJ, Jallerat Q, Palchesko RN, Park JH, Grodzicki MS, Shue H-J, Ramadan MH, Hudson AR, Feinberg AW (2015) Three-dimensional printing of complex biological structures by freeform reversible embedding of suspended hydrogels. Sci Adv 1(9): e1500758
6. Kang H-W, Lee SJ, Ko IK, Kengla C, Yoo JJ, Atala A (2016) A 3D bioprinting system to produce human-scale tissue constructs with structural integrity. Nat Biotechnol 34:312–319
7. Ragelle H, Tibbitt MW, Wu S-Y, Castillo MA, Cheng GZ, Gangadharan SP, Anderson DG, Cima MJ, Langer R (2018) Surface tension-assisted additive manufacturing. Nat Commun 9(1):1184
8. Bücking TM, Hill ER, Robertson JL, Maneas E, Plumb AA, Nikitichev DI (2017) From medical imaging data to 3D printed anatomical models. PLoS One 12(5):e0178540
9. Sun W, Starly B, Nam J, Darling A (2005) Bio-CAD modeling and its applications in computer-aided tissue engineering. Comput Aided Des 37(11):1097–1114
10. Loessner D, Meinert C, Kaemmerer E, Martine LC, Yue K, Levett PA, Klein TJ, Melchels FPW, Khademhosseini A, Hutmacher DW (2016) Functionalization, preparation and use of cell-laden gelatin methacryloyl–based hydrogels as modular tissue culture platforms. Nat Protoc 11:727–746
11. Fairbanks BD, Schwartz MP, Bowman CN, Anseth KS (2009) Photoinitiated polymerization of PEG-diacrylate with lithium phenyl-2,4,6-trimethylbenzoylphosphinate: polymerization rate and cytocompatibility. Biomaterials 30(35):6702–6707

Part IV

Applicative Scenarios

Chapter 14

Bioprinting of Complex Vascularized Tissues

Wei Zhu, Claire Yu, Bingjie Sun, and Shaochen Chen

Abstract

Functional vasculature is crucial for the maintenance of living tissues via the transport of oxygen, nutrients, and metabolic waste products. As a result, insufficient vascularization in thick engineered tissues will lead to cell death and necrosis due to mass transport and diffusional constraints. To circumvent these limitations, we describe the development of a microscale continuous optical bioprinting (μCOB) platform for 3D printing complex vascularized tissues with superior resolution and speed. By using the μCOB system, endothelial cells and other supportive cells can be printed directly into hydrogels with precisely controlled distribution and subsequent formation of lumen-like structures in vitro.

Key words 3D bioprinting, Vasculature, Complex microarchitecture, Hydrogels, Tissue engineering

1 Introduction

Tissue engineering is an emerging field of research that aims at developing biological substitutes that can restore, maintain, or improve tissue function or a whole organ [1, 2]. Ultimately, such engineered tissue substitutes or organs can be used for in vivo transplantation or in vitro drug screening. One of the key bottlenecks in tissue engineering is to provide sufficient vascularization within thick engineered tissue constructs. Without proximity (~150–200 μm) to the vascular network, cellular viability and tissue function will be compromised within a very short time [3, 4]. The incorporation of angiogenic growth factors is a typical strategy to stimulate the host vasculature to infiltrate the engineered tissue constructs after implantation [5–7]. However, this approach is costly and not efficacious due to the relatively slow revascularization process and short half-life of growth factors in vivo. A growing and promising strategy is with the formation of prevascularized engineered tissue constructs in vitro prior to implantation. Currently, this has shown great potential in providing sufficient vascularization as well as improving cellular activity and tissue functions in vivo [8–11]. PDMS molding techniques have also been used to fabricate

Alberto Rainer and Lorenzo Moroni (eds.), *Computer-Aided Tissue Engineering: Methods and Protocols*, Methods in Molecular Biology, vol. 2147, https://doi.org/10.1007/978-1-0716-0611-7_14, © Springer Science+Business Media, LLC, part of Springer Nature 2021

spatially defined endothelial cords, which significantly improved the speed and extent of vascularizing the engineered tissue after implantation compared to randomly seeded endothelial cells [4]. However, such molding techniques are limited to fabricating tissues with simple designs and new physical molds are needed if the design is to be altered or adjusted, which can be costly and time-consuming. 3D bioprinting has emerged as a promising solution for building complex tissue constructs with excellent flexibility and versatility. In particular, nozzle-based bioprinters have been successfully adopted to print vascularized 3D tissues indirectly with sacrificial inks [3, 12–14] or directly with endothelial cells [15, 16]. Nozzle-based bioprinters deposit the bioink line by line in a serial manner that can be relatively slow for large-scale tissues. Furthermore, the interfaces between the lines can compromise the mechanical integrity of the printed tissues. In light of these technological challenges, optical projection printing has arisen as the next-generation microfabrication technique with superior speed, resolution, and flexibility to fabricate complex tissue constructs.

Here, we present a rapid optical projection printing platform—microscale continuous optical bioprinting (μCOB)—for creating complex vascularized tissues directly with endothelial cells and other supportive cells in vitro [17]. These vascularized tissues were printed with unprecedented speed and resolution, featuring complex microarchitectures and precisely controlled cell distribution. We demonstrated that the printed vascularized constructs exhibited endothelial networks with lumen-like structures after 1-week culture in vitro.

2 Materials

2.1 General Equipment

1. Heated magnetic stirrer.
2. Freeze dryer.
3. Laboratory fume hood.
4. Class 2 biosafety cabinet.
5. Water bath (37 °C).
6. −80 °C freezer

2.2 Glycidyl Methacrylate-Hyaluronic Acid (GM-HA) Synthesis

1. Hyaluronic acid (MW 200 kDa).
2. Acetone.
3. Deionized (DI) water.
4. Triethylamine (TEA).
5. Glycidyl methacrylate (GM).
6. Dialysis tubing (MWCO 3.5 kDa).

2.3 Gelatin Methacrylate (GelMA) Synthesis

1. Porcine skin gelatin.
2. Dulbecco's phosphate-buffered saline (DPBS).
3. Methacrylate anhydride (MA).
4. DI water.
5. Dialysis tubing (MWCO 13.5 kDa).

2.4 Photoinitiator Lithium Phenyl-2,4,6 Trimethyl-benzoylphosphinate (LAP) Synthesis

1. 2,4,6-trimethylbenzoyl chloride.
2. Dimethyl phenylphosphonite.
3. Lithium bromide.
4. 2-butanone.

2.5 Cell and Tissue Cultures

1. Human umbilical vein endothelial cells (HUVECs).
2. Endothelial cell growth medium (EGM-2, Lonza).
3. C3H/10T1/2 cells (10T1/2 s).
4. Dulbecco's modified eagle medium (DMEM).
5. Fetal bovine serum (FBS).
6. 0.05% trypsin-EDTA
7. 0.25% trypsin-EDTA
8. DPBS.

2.6 μCOB Setup

1. Digital micromirror array device (DMD, DLP9500UV, Texas Instrument).
2. UV LED light source (365 nm, Hamamatsu).
3. 3-axis linear stage with motion controller (Newport).
4. Projection optics.
5. Motorized syringe pump.
6. Computer.

2.7 Cell Viability Assay

1. LIVE/DEAD® Viability/Cytotoxicity Kit (Invitrogen).
2. DPBS.

2.8 Immuno-fluorescence Staining

1. Fixation buffer: 4% Paraformaldehyde (PFA) in DPBS.
2. Blocking/permeabilization buffer: 2% bovine serum albumin (BSA), 0.1% Triton X-100 in DPBS.
3. Anti-CD31 primary antibody: goat anti-human PECAM-1 (M-20, sc-1506, Santa Cruz Biotechnology); 1:100 in blocking/permeabilization buffer. Prepare shortly before use.
4. Anti-alpha-smooth muscle actin primary antibody: rabbit anti-mouse α-SMA (ab5694, Abcam); 1:100 in blocking/permeabilization buffer. Prepare shortly before use.
5. CF488A donkey anti-rabbit IgG (H+L) (20015, Biotium).
6. CF568 donkey anti-goat IgG (H+L) (20106, Biotium).

3   Methods

3.1   GM-HA Synthesis

1. Dissolve 1 g of hyaluronic acid in 100 mL of acetone/water (50/50) solution overnight to prepare a 1% w/v solution.

2. Add 7.2 mL TEA (20-fold molar excess) until thoroughly mixed (*see* **Note 1**).

3. Add 7.2 mL GM (20-fold molar excess) until thoroughly mixed (*see* **Note 2**).

4. Stir overnight at room temperature (*see* **Note 3**).

5. Dialyze the resulting solution against DI water with 3.5 kDa tubing at room temperature for 48 h (*see* **Note 4**).

6. Collect and freeze the dialyzed solution overnight at −80 °C (*see* **Note 5**).

7. Lyophilize the frozen solution for 48 h at 0.04 mbar and −50 °C.

8. Store the lyophilized GM-HA at −80 °C for future use.

3.2   GelMA Synthesis

1. Dissolve 10 g of porcine skin gelatin in 100 mL DPBS by stirring at 60 °C to prepare a 10% w/v solution (*see* **Note 6**).

2. Add 8 mL of MA to the solution at a rate of 0.5 mL/min (*see* **Note 7**).

3. Allow the reaction to continue for 3 h at 60 °C with constant stirring.

4. Add an equivalent volume of warm DPBS to dilute the resulting solution (*see* **Note 8**).

5. Dialyze the diluted solution against DI water with 13.5 kDa tubing for 1 week at 40 °C (*see* **Note 9**).

6. Collect and freeze the dialyzed solution overnight at −80 °C (*see* **Note 5**).

7. Lyophilize the frozen solution for 48 h at 0.04 mbar and −50 °C.

8. Store the lyophilized GelMA at −80 °C for future use.

3.3   LAP Synthesis

1. All the following steps are performed in a chemical fume hood.

2. Add 3.2 g of 2,4,6-trimethylbenzoyl chloride dropwise to an equal molar amount of dimethyl phenylphosphonite (3 g) with constant stirring under argon.

3. Allow the reaction to continue for 18 h.

4. Dissolve 6.1 g lithium bromide to 100 mL of 2-butanone and add into the previous mixture at fourfold excess.

5. Heat the reaction to 50 °C until a solid precipitate forms in about 10 min.

6. Allow the mixture to cool down to room temperature and rest overnight.

7. Filter the resulting mixture and use 2-butanone to wash the filtrate three times to remove the unreacted lithium bromide.

8. Remove the excess solvent by vacuum and leave LAP in a white solid chunk state.

9. Pestle the chunk LAP into powder and store it at $-80\,^{\circ}\text{C}$ under argon for future use.

3.4   Cell and Tissue Cultures

1. Culture HUVECs in EGM-2 according to the protocol from the vendor (*see* **Note 10**).

2. Culture 10T1/2 s in DMEM supplemented with 10% FBS according to the protocol from the vendor.

3. Culture the bioprinted tissue constructs in EGM-2 (*see* **Note 11**).

3.5   μCOB System Setup

1. The schematic of the μCOB is shown in Fig. 1.

2. The DMD chip is used to dictate the optical pattern projected onto the prepolymer solution on the fabrication stage.

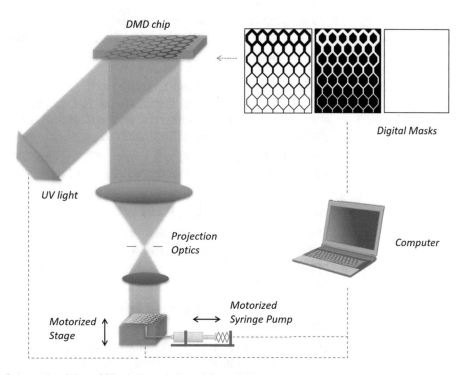

Fig. 1 Schematic of the μCOB platform (adapted from [17])

3. A UV light source (365 nm) is used to induce the photopolymerization. A set of optics are placed in front of the UV light source to collimate and project the UV light to the DMD chip. It is recommended to illuminate the entire DMD chip with even light intensity. A homogenizer can be used to modulate the light intensity.

4. A set of projection optics are placed below the DMD chip to project the digital patterns to the fabrication reservoir. The magnification of the projection optics dictates the resolution and overall dimension of the printed structures.

5. A motorized syringe pump system is connected to the fabrication reservoir to add and remove the prepolymer solution.

6. A computer is used to control the DMD, the UV light source, the syringe pump system, and the motorized stage. A software is developed to synchronize the control of all the components.

7. 3D models can be built in computer-aided-design (CAD) software or from computed tomography (CT) and magnetic resonance imaging (MRI) scans.

8. The digital masks are sliced from the 3D models and fed into the DMD continuously to alter the optical pattern projected onto the prepolymer solution.

9. 3D printing can be realized by simultaneously moving the stage and updating the digital masks which are both controlled by the computer.

3.6 3D Bioprinting of Vascularized Tissues

1. Design a set of digital masks according to the example provided in Fig. 1, suitable for the fabrication of tissues with gradient channel widths (ranging from 50 μm to 250 μm), mimicking the branching structure of a vasculature network.

2. Digest HUVECs and 10T1/2 s by 0.05% trypsin-EDTA and 0.25% trypsin-EDTA, respectively (*see* **Note 12**).

3. Mix these two cell types at a ratio of 50:1 (40 million/mL HUVECs and 800,000/mL 10T1/2 s) (*see* **Note 13**).

4. Prepare prepolymer A by dissolving 5% (w/v) GelMA and 0.15% (w/v) LAP in DPBS (*see* **Note 14**).

5. Prepare prepolymer B by dissolving 5% (w/v) GelMA, 2% (w/v) GM-HA, and 0.15% (w/v) LAP in DPBS.

6. Filter the prepolymer A and B solutions with a 0.22 μm sterile filter.

7. Load prepolymer A onto the fabrication stage and the first mask on the left in Fig. 1 onto the DMD chip.

8. Expose the UV light for 15 s to fabricate the base layer. Remove the unpolymerized part of the prepolymer A and wash with DPBS.

9. Mix prepolymer B and cell solution at a volume ratio of 1:1 to get a final composition of 2.5% (w/v) GelMA, 1%(w/v) GM-HA, 0.15%(w/v) LAP, 20 million HUVECs/mL, and 40,000 10T1/2 s/mL (*see* **Note 15**).

10. Load the cell-prepolymer mixture on top of the base layer and the second mask in the middle in Fig. 1 onto the DMD chip.

11. Expose the UV light for 15 s to print the vascular layer. Remove the unpolymerized part of the cell-prepolymer mixture and wash with DPBS.

12. Load prepolymer A onto the fabrication stage and the third mask on the right in Fig. 1 onto the DMD chip.

13. Expose the UV light for 15 s to print the top layer to enclose the vascular network. Remove the non-polymerized part of the cell-prepolymer mixture and wash with DPBS.

14. Transfer the printed vascularized tissue construct to a well plate and culture it with EGM-2.

3.7 Cell Viability Test

1. Cell viability is performed on day 1, day 3, and day 7 after printing.

2. Prepare the live/dead assay by mixing a final concentration of 2 μM calcein AM (live cell stain) and 4 μM ethidium homodimer-1 (dead cell stain) solution with DPBS and vortex to mix thoroughly (*see* **Note 16**).

3. Remove the culture medium and wash the tissue construct with DPBS three times (*see* **Note 17**).

4. Add the live/dead working solution to the tissue construct and incubate at room temperature for 30 min (*see* **Note 18**).

5. Remove the live/dead assay after the incubation and add DPBS to the tissue construct.

6. Take fluorescence images of the tissue construct at different heights (Z positions) (*see* **Note 19**).

7. Cell viability can be calculated as the ratio between live cells and total cells (*see* **Note 20**).

3.8 Immuno-fluorescence Staining

1. Remove the culture media and fix the tissue construct immediately with fixation buffer for 30 min at room temperature.

2. Remove the fixation buffer and rinse the tissue construct with DPBS three times for 5 min each.

3. Treat the samples with blocking and permeabilization buffer for 1 h at room temperature.

4. Near the end of the blocking and permeabilization process, prepare primary antibody solution by diluting CD31 and α-SMA in the 2% BSA and 0.1% Triton X-100 solution both at a ratio of 1:100.

5. Remove the blocking and permeabilization buffer and apply the primary antibody solution to the tissue construct.

6. Incubate the tissue construct in the primary antibody solution overnight at 4 °C.

7. After incubation with the primary antibodies, rinse the tissue construct three times with DPBS for 5 min each.

8. Incubate the tissue construct in the diluted fluorochrome-conjugated secondary antibodies for 1 h at room temperature in the dark.

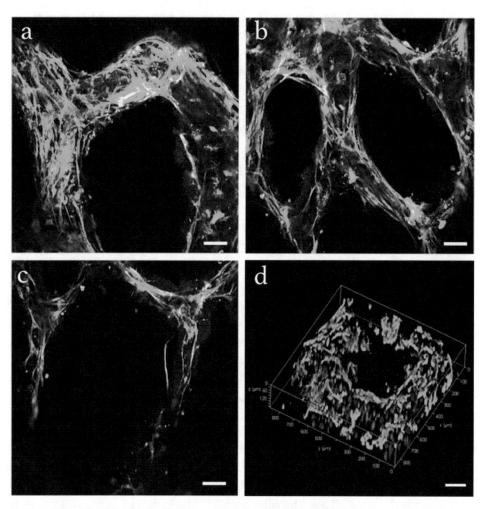

Fig. 2 Immunofluorescence staining of the endothelial network formation after 1-week culture in vitro (adapted from [17]). (**a–c**) Fluorescent microscopy images show HUVECs (Green, CD31-positive) and supportive mesenchymal cells (10T1/2, Purple, α-SMA-positive) aligned within the patterned gradient channel regions with different vessel sizes. (**d**) 3D view of the endothelial cells lining along the printed microchannel walls by confocal microscopy. Endothelial cells are labeled by fluorescent cell tracker (red) and stained by CD31 (green). Scale bars: 100 μm

9. Remove the secondary antibody solution and rinse the tissue construct with DPBS three times with DPBS for 5 min each.

10. Image the tissue construct with confocal microscopy (Fig. 2).

4 Notes

1. Add the TEA dropwise with a syringe to make sure it is thoroughly mixed.

2. Add the GM dropwise with a syringe to make sure it is thoroughly mixed.

3. Cover the reaction solution from light with aluminum foil.

4. Keep stirring the DI water during dialysis. Change the DI water after 2 h, 4 h, and 24 h.

5. Freeze the solution at an angle to maximize surface area for subsequent lyophilization.

6. Use a water bath to control the temperature at 60 °C and allow the gelatin to completely dissolve before proceeding to the next step.

7. Fill the reaction container with argon during the reaction.

8. Warm the DPBS to 50 °C prior to the end of the 3 h.

9. Keep stirring the DI water during dialysis. Change the DI water 2–3 times a day. Heat the fresh DI water to 40 °C before the change.

10. Use passage 3–6 HUVECs for bioprinting.

11. Change the medium every other day and use a 1 mL pipette instead of the vacuum line to remove the old medium.

12. Incubate the cells in trypsin-EDTA solution for 5 min at 37 °C in the CO_2 incubator.

13. The cell concentrations here are twice the final concentrations. This cell solution will be mixed with the prepolymer solution at a volume ratio of 1:1.

14. Dissolve GelMA by heating the solution at 37 °C. Do not vortex the GelMA solution to avoid bubbles.

15. Gently pipette the mixture of the prepolymer and the cells to make a homogenous solution and avoid creating bubbles.

16. The live dead working solution should be used within 1 day, since the calcein AM aqueous solution is susceptible to hydrolysis.

17. Serum esterase present in the serum-supplemented media can hydrolyze the calcein AM and cause increase in extracellular fluorescence. Therefore, it is important to wash the tissue constructs well and gently with 500–1000 volumes of DPBS.

18. Protect the plate from light with aluminum foil. To facilitate diffusion of the working solution into the tissue construct, it can also be incubated at 37 °C or for a longer incubation time.

19. The excitation and emission wavelengths for calcein AM are ~495 nm/~515 nm. The excitation and emission wavelengths for ethidium homodimer-1 are ~495 nm/~635 nm.

20. ImageJ or Fiji software can be used for cell quantification automatically or manually. Expected viability values are in the range of 85–95%.

References

1. Langer R, Vacanti JP (1993) Tissue engineering. Science 260:920–926. https://doi.org/10.1126/science.8493529

2. Khademhosseini A, Langer R, Borenstein J, Vacanti JP (2006) Microscale technologies for tissue engineering and biology. Proc Natl Acad Sci 103:2480–2487. https://doi.org/10.1073/pnas.0507681102

3. Kolesky DB, Truby RL, Gladman AS, Busbee TA, Homan KA, Lewis JA (2014) 3D bioprinting of vascularized, heterogeneous cell-laden tissue constructs. Adv Mater 26:3124–3130. https://doi.org/10.1002/adma.201305506

4. Baranski JD, Chaturvedi RR, Stevens KR, Eyckmans J, Carvalho B, Solorzano RD, Yang MT, Miller JS, Bhatia SN, Chen CS (2013) Geometric control of vascular networks to enhance engineered tissue integration and function. Proc Natl Acad Sci 110:7586–7591. https://doi.org/10.1073/pnas.1217796110

5. Lee KY, Peters MC, Anderson KW, Mooney DJ (2000) Controlled growth factor release from synthetic extracellular matrices. Nature 408:998–1000. https://doi.org/10.1038/35050141

6. Fischbach C, Mooney DJ (2007) Polymers for pro- and anti-angiogenic therapy. Biomaterials 28:2069–2076. https://doi.org/10.1016/j.biomaterials.2006.12.029

7. Richardson TP, Peters MC, Ennett AB, Mooney DJ (2001) Polymeric system for dual growth factor delivery. Nat Biotechnol 19:1029–1034. https://doi.org/10.1038/nbt1101-1029

8. Levenberg S, Rouwkema J, Macdonald M, Garfein ES, Kohane DS, Darland DC, Marini R, van Blitterswijk CA, Mulligan RC, D'Amore PA, Langer R (2005) Engineering vascularized skeletal muscle tissue. Nat Biotechnol 23:879–884. https://doi.org/10.1038/nbt1109

9. Caspi O, Lesman A, Basevitch Y, Gepstein A, Arbel G, Huber I, Habib M, Gepstein L, Levenberg S (2007) Tissue engineering of vascularized cardiac muscle from human embryonic stem cells. Circ Res 100:263–272. https://doi.org/10.1161/01.RES.0000257776.05673.ff

10. Koike N, Fukumura D, Gralla O, Au P, Schechner JS, Jain RK (2004) Tissue engineering: creation of long-lasting blood vessels. Nature 428:138–139. https://doi.org/10.1038/428138a

11. Chen X, Aledia AS, Ghajar CM, Griffith CK, Putnam AJ, Hughes CCW, George SC (2009) Prevascularization of a fibrin-based tissue construct accelerates the formation of functional anastomosis with host vasculature. Tissue Eng Part A 15:1363–1371. https://doi.org/10.1089/ten.tea.2008.0314

12. Bertassoni LE, Cecconi M, Manoharan V, Nikkhah M, Hjortnaes J, Cristino AL, Barabaschi G, Demarchi D, Dokmeci MR, Yang Y, Khademhosseini A (2014) Hydrogel bioprinted microchannel networks for vascularization of tissue engineering constructs. Lab Chip 14:2202–2211. https://doi.org/10.1039/c4lc00030g

13. Kolesky DB, Homan KA, Skylar-Scott MA, Lewis JA (2016) Three-dimensional bioprinting of thick vascularized tissues. Proc Natl Acad Sci U S A 113:3179–3184. https://doi.org/10.1073/pnas.1521342113

14. Miller JS, Stevens KR, Yang MT, Baker BM, Nguyen D-HT, Cohen DM, Toro E, A a C, P a G, Yu X, Chaturvedi R, Bhatia SN, Chen CS (2012) Rapid casting of patterned vascular networks for perfusable engineered three-dimensional tissues. Nat Mater 11:768–774. https://doi.org/10.1038/nmat3357

15. Jia W, Gungor-Ozkerim PS, Zhang YS, Yue K, Zhu K, Liu W, Pi Q, Byambaa B, Dokmeci MR, Shin SR, Khademhosseini A (2016) Direct 3D

bioprinting of perfusable vascular constructs using a blend bioink. Biomaterials 106:58–68. https://doi.org/10.1016/j.biomaterials. 2016.07.038

16. Zhang YS, Arneri A, Bersini S, Shin SR, Zhu K, Goli-Malekabadi Z, Aleman J, Colosi C, Busignani F, Dell'Erba V, Bishop C, Shupe T, Demarchi D, Moretti M, Rasponi M, Dokmeci MR, Atala A, Khademhosseini A (2016) Bioprinting 3D microfibrous scaffolds for engineering endothelialized myocardium and heart-on-a-chip. Biomaterials 110:45–59. https://doi.org/10.1016/j.biomaterials. 2016.09.003

17. Zhu W, Qu X, Zhu J, Ma X, Patel S, Liu J, Wang P, Lai CSE, Gou M, Xu Y, Zhang K, Chen S (2017) Direct 3D bioprinting of prevascularized tissue constructs with complex microarchitecture. Biomaterials 124:106–115. https://doi.org/10.1016/j.biomaterials. 2017.01.042

Chapter 15

A Scaffold Free 3D Bioprinted Cartilage Model for In Vitro Toxicology

Pallab Datta, Yang Wu, Yin Yu, Kazim K. Moncal, and Ibrahim T. Ozbolat

Abstract

Bioprinting has emerged as a promising method for precise spatiotemporal patterning of biological materials such as living cells, genetic materials, and proteins, which are sensitive to any other fabrication techniques. Bioprinting allows the generation of tissue constructs and models that closely mimic the anatomical and physiological attributes of a chosen tissue. In vitro toxicology assays can greatly benefit from bioprinting as drugs can be screened with higher efficiencies in a significantly reduced period. This protocol describes a method for fabricating bioprinted cartilage constructs which can be used for in vitro toxicology studies employing a scalable "tissue strand" bioprinting modality.

Key words Scaffold-free bioprinting, Tissue strands, Micro-tissue fabrication

1 Introduction

Bioprinting, techniques to deposit cells and biological materials into predefined three-dimensional (3D) patterns, can be carried out in scaffold-based or scaffold-free methods [1]. The scaffold-based method makes use of a hydrogel/polymer matrix which is biocompatible. Some limitations associated with scaffold-based techniques are biomaterials for support structure, number of cells that can be incorporated, slow cell growth kinetics, and material degradation [2, 3]. In comparison, scaffold-free techniques allow cell bioprinting at high concentrations approaching cell concentrations in native tissue. Also, scaffold-free bioprinting does not include exogenous biomaterials, which allows for more space for ECM deposition, better cell-cell interactions, generation of tissues with close biomimicry, preservation of cell functionality for longer term, and elimination of biodegradation issues [4].

Bioprinting can be accomplished by any of three principal methods, namely extrusion-based (EBB), droplet-based (DBB),

Alberto Rainer and Lorenzo Moroni (eds.), *Computer-Aided Tissue Engineering: Methods and Protocols*,
Methods in Molecular Biology, vol. 2147, https://doi.org/10.1007/978-1-0716-0611-7_15,
© Springer Science+Business Media, LLC, part of Springer Nature 2021

or laser-based (LBB) bioprinting [5, 6]. EBB uses pneumatic or mechanical extrusion to print the cell suspension in the required pattern. DBB, which ejects the cell-laden droplets, is mediated by piezoelectric, acoustic, thermal, or electrostatic actuation [7]. In LBB, a laser energy-absorbing material evaporates after irradiation, followed by deposition on a layer of biological ink. EBB is preferred to fabricate mechanically enhanced tissue constructs, although DBB and LBB offer greater precision. One concern of bioprinting is the cell viability after printing. Shear stresses in EBB, thermal stresses in DBB, and irradiation energy in LBB are the major causes of cell death in bioprinted constructs [8].

Despite the limitations, bioprinted constructs can be an important tool in preclinical in vitro testing of pharmaceuticals [9]. Certain drugs can cause tissue-specific toxicity, which can be investigated at the early stages of development. Drug toxicity has been routinely assessed in critical organs or tissues of the body, especially liver [10]. An increase in antibiotic-resistant bacterial strains has driven the development of new antibiotics; however, chondrotoxicity is one of the major concerns for the safe use of antibiotics. Bioprinted chondrocyte constructs can be used in the preclinical phase to determine the toxic effects of a drug on cartilage tissue [4].

In this chapter, a novel method for microfabrication of scalable tissue strands as building units will be presented. Cell aggregation and fusion are achieved without the use of liquid printing medium or an external mold. The scaffold-free cartilage tissue model presented in this work closely recapitulates the articular cartilage biology and physiology to enable the use of such tissues in toxicology studies in vitro.

2   Materials

All solutions were prepared using ultrapure water and analytical grade reagents.

2.1 Sodium Alginate Preparation

1. Sodium alginate solution: powder treated with ultraviolet (UV) light for sterilization with three 30-min cycles, 4% w/v in sterilize deionized water, stir overnight in a magnetic stirrer.

2. $CaCl_2$ solution: powder treated with UV light for sterilization with three 30-min cycles, 4% w/v in sterilize deionized water (see Note 1).

3. Sodium citrate solution: powder treated with UV light for sterilization with three 30-min cycles, 4% w/v in sterilize deionized water.

2.2   Cell Preparation

1. Articular cartilage from femur condyle of young adult bovine.

2. Hank's Balanced Salt Solution (HBSS) buffer: supplemented with 100 U/μL penicillin, 100 μg/mL streptomycin, and 2.5 μg/mL fungizone.

3. Culture medium: Dulbecco's modified culture medium (DMEM) and Ham's F12 medium (1:1) supplemented with 10% fetal bovine serum, 50 μg/mL ʟ-ascorbate, 100 U/mL penicillin, 100 μg/mL streptomycin, and 2.5 μg/mL fungizone

4. Digestion buffer: 0.25 mg/mL Collagenase type I and 0.25 mg/mL Pronase E in culture medium.

2.3   Bioprinting Equipment

1. Pneumatic air dispenser.

2. Mechanical pump.

3. Coaxial needle with a 22 G inner nozzle (0.71 mm and 0.41 mm for outer and inner diameters, respectively) and a 14 G outer nozzle (2.11 mm and 1.69 mm for outer and inner diameters, respectively) (Fig. 1a) (*see* **Note 2**).

2.4   Cell Viability Assay

1. Calcein/ethidium solution: 1.0 mM Calcein acetoxymethylester (calcein AM) and 1.0 mM ethidiumhomodimer-2 in phosphate-buffered saline (PBS).

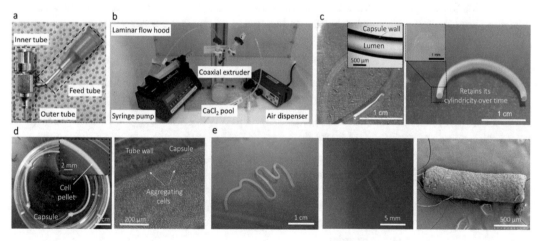

Fig. 1 Fabrication of tissue strands. (**a**) A home-made coaxial nozzle apparatus. (**b**) Setup for co-axial extrusion of alginate capsules. (**c**) Left: Alginate capsules with well-defined tubular morphology. Right: Capsules preserved their cylindrical morphology without any collapse in culture over time. (**d**) Left: Microinjected cell pellet inside alginate capsules supporting aggregation in a few days. Right: Aggregation began around the luminal surface of the capsule. (**e**) Left: A nearly 8 cm long tissue strand after decrosslinking the capsule. Middle: Tissue strands were self-assembled into a "T" shape. Right: Well-defined cylindrical morphology was achieved as shown in an SEM image

2.5 Histology and Immunohisto-chemistry Analysis	1. Paraformaldehyde (PFA): 4% in PBS.
	2. Hematoxylin.
	3. Safranin O-fast green.
	4. Primary antibody: rabbit anti-human polyclonal antibodies against collagen type II and aggrecan.
	5. Secondary antibody: goat anti-rabbit antibody.
	6. Vectastain ABC kit.
	7. DAB Peroxidase Substrate Kit.
2.6 DMMB Assay for sGAG Content Evaluation	1. 1,9-dimethylmethylene blue (DMMB) solution at a concentration of 16 μg/mL.
	2. Quant-iT PicoGreen dsDNA Assay Kit.
2.7 Total RNA Extraction, Reverse Transcription, and Real-Time PCR Analysis	1. TRIzol reagent.
	2. RNeasy Mini Kit.
	3. TaqMan Micro RNA reverse transcription kits.
	4. SYBR Green Real-Time PCR kit.

3   Methods

3.1 Preparation of Alginate Capsule	1. Load 4% sodium alginate into a 5 mL syringe, and connect the syringe to the feed tube of the coaxial nozzle.
	2. Load 4% $CaCl_2$ into another 5 mL syringe, and connect the syringe to the inner tube of the coaxial nozzle.
	3. Print the alginate at the air pressure of 82.7 kPa, and the $CaCl_2$ at a feed rate of 16 mL/min (Fig. 1b) (*see* **Note 3**).
	4. Collect the pre-crosslinked alginate capsules in a $CaCl_2$ pool (*see* **Note 4**).
	5. Leave the alginate capsules in $CaCl_2$ solution overnight for complete crosslinking (Fig. 1c).
3.2 Cell Preparation	1. Rinse the articular cartilage in HBSS buffer with 100 U/μL penicillin, 100 μg/mL streptomycin, and 2.5 μg/mL fungizone.
	2. Mince the cartilage into fine pieces.
	3. Digest the cartilage pieces overnight in digestion buffer.
	4. Culture the isolated cells in growth medium at 37 °C, 5% CO_2.
	5. Fabrication of tissue strands.
	6. Expand the chondrocytes to approximately 100 million cells.
	7. Suspend the cells in culture medium in a 50 mL conical tube.

8. Centrifuge the cell suspension at 2000 rpm for 10 min to obtain the cell pellet. Discard supernatant (*see* **Note 5**).

9. Aspirate the cell pellet into the syringe unit (*see* **Notes 6** and **7**).

10. Inject the cell pellet into the fabricated alginate tubular constructs (*see* **Note 8**).

11. Incubate the pellets for 7 days (Fig. 1d).

12. Dissolve the alginate tube in sodium citrate solution for 5 min to obtain the pure cartilage strands (Fig. 1e).

13. Culture tissue strands in medium as Subheading 3.3, **step 4**.

3.3 Cell Viability Assay

1. Rinse the strand with HBSS twice, 15 min each time.

2. Place the strand into calcein/ethidium solution.

3. Incubate the strand for 30 min at 37 °C with 5% CO_2.

4. Remove the calcein/ethidium solution.

5. Rinse the strand with HBSS twice, 15 min each time.

6. Image the strand using a laser scanning confocal microscope (LSCM), with a depth of 1000 μm at 20 μm intervals for Z-axis projections.

7. Apply ImageJ software for intensity quantification of the red- and green-stained tissue strands.

3.4 Histology Analysis

1. Fix cells in the strand with PFA.

2. Freeze the strand, and cut it into 10 μm sections.

3. Perform hematoxylin and Safranin O-fast green staining according to standard protocols [11].

3.5 Immunohistochemistry Analysis

1. Prepare the tissue sections as Subheadings 3.6, **steps 1** and **2**.

2. Incubate the section for 30 min in blocking solution to prevent nonspecific binding.

3. Incubate the section with rabbit anti-human polyclonal antibodies against collagen type II and aggrecan overnight at room temperature.

4. Use the goat anti-rabbit secondary antibody for detection.

5. Visualize the reaction products by the Vectastain ABC kit and the DAB peroxidase substrate kit according to the manufacturer's instructions.

6. Omit the primary antibodies to generate negative controls.

7. Observe the section under a microscope.

3.6 DMMB Assay for sGAG Content Evaluation

1. Dilute the cell lysates to contain 4×10^3 cells/μL in 50 μL, and add 150 μL 1,9-dimethylmethylene blue (DMMB) dye solution (*see* **Note 9**).

2. Measure the absorbance at 530 nm using a microplate reader.

3. Digest the 2-week-cultured tissue strands as well as the native articular cartilage in the papain buffer.

4. Apply a Quant-iTTM PicoGreen dsDNA Assay Kit according to the manufacturer's instructions (*see* **Note 10**).

5. Measure the fluorescence intensity by a microplate reader using a 480 nm excitation and 520 nm emission wavelength.

6. Normalize the sulfated glycosaminoglycans (sGAG) content from each sample to dsDNA content; results are presented as sGAG content per cell.

3.7 Total RNA Extraction, Reverse Transcription, and Real-Time PCR Analysis

1. Homogenize tissue strands in TRIzol reagent.

2. Extract total RNA using the RNeasy Mini Kit according to the manufacturer's instructions.

3. Reverse transcribe cDNA using TaqMan Micro RNA reverse transcription kits according to instructions from the vendor.

4. Apply SYBR Green Real-Time PCR kit to analyze the transcription levels of cartilage-matrix-related genes, including collagen type II, aggrecan, and chondrogenic transcription factor Sox9.

3.8 Bioprinting of Tissue Strands

1. Load tissue strand within capsules into the detachable nozzle (Fig. 2a).

2. Add 4% sodium citrate solution onto the capsules to decrosslink the alginate tube.

3. Remove excess sodium citrate solution from the remaining tissue strand.

4. Mount the nozzle with tissue strand along with the mechanical dispensing system onto the bioprinter (Fig. 2a).

5. Place a filter paper hydrated with cell culture media on the printing platform.

6. Print the strand using printing speed at 100 mm/min and an extrusion speed at 50.8 mm/min.

7. Load a new strand and repeat the above processes until the entire structure is printed. Raise z axis 0.5 mm for the printing of additive layer.

8. Hydrate the tissue strands for sufficient time to allow tissue fusion, and transfer them to the tissue culture dish.

3.9 Self-Assembly of Tissue Strands

1. Place multiple constructs in close proximity to each other in a 150 mm Petri dish.

2. Supplement a minimum amount of culture media to ensure cells survive without loss of contact (Fig. 2b).

3. Assess the self-assembly of strands via imaging to monitor the fusion process with minimal disturbance.

a

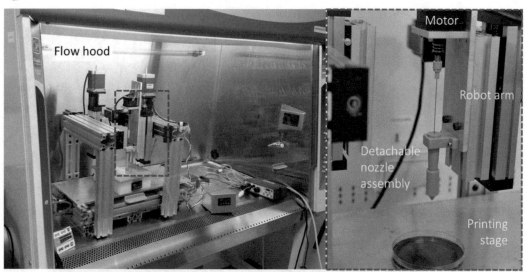

b

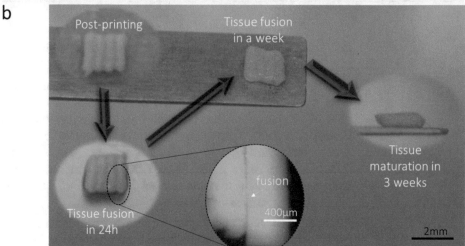

Fig. 2 (**a**) Images of printed tissue morphology over 3 weeks of incubation showing the fusion of printed strands. (**b**) The printer and print head design loaded with detachable nozzle assembly for tissue strand printing

3.10 Compression Tests on Printed Tissue

1. Measure the thickness of the printed tissue by a laser measurement system.

2. Compress the tissue to 20% strain at 2 mm/s with a 10 N load held for 5 min using a mechanical testing machine (Fig. 3).

3. Calculate the Young's modulus during the initial loading period (between 15% and 20% strain).

4. Harvest the native cartilage samples from the trochlear groove of bovine femur condyles and perform compression analysis for comparison.

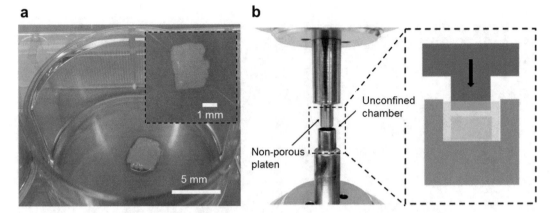

Fig. 3 (**a**) 3D Bioprinted cartilage tissue samples 3 weeks post-bioprinting were trimmed (dashed inset) to fit into the device to test Young's modulus (for compression). (**b**) Apparatus (a materials testing machine) and the scheme (dashed inset) for the stress-relaxation test

4   Notes

1. PBS should not be used to CaCl$_2$ solution as precipitation of later is often observed.

2. Coaxial nozzle consists of feed tube, outer tube, and inner tube. Create a hole with the same outer diameter as the feed tube in the barrel of the outer tube to attach the feed tube. Remove the luer lock hub on the barrel of the outer tube using a lathe and grind the tip to ensure the inner and the outer dispensing tips were even. Align the inner and outer tubes concentrically using a stainless steel fixture manufactured by micro-milling, and assemble using laser welding.

3. In order to prevent nozzle clogging caused by alginate, which may crosslink and accumulate at the tip of the nozzle, operator could turn on the syringe pump to keep the CaCl$_2$ running prior to cell-loaded alginate.

4. Occasionally, the alginate solution may form a big droplet at the nozzle tip without generating continuous conduit. Operator could use a sterilized tweezer to draw the alginate droplet to initiate conduit extrusion.

5. Supernatant should be removed as much as possible. Operators could use 1 mL, 200 mL, and 10 mL pipettes to remove the supernatant sequentially.

6. Cell pellet should be aspirated as slow as possible to avoid bubbles inside the syringe unit.

7. Aspirated cell pellet can be kept in the syringe unit for up to half an hour to settle down by gravity, in case of bubble generation. Keep the entire cell-loaded syringe unit in the incubator.

8. Inject cell pellet as slowly as possible to prevent leakage of cells from the alginate capsule.

9. DNA is known to interfere with DMMB assay. Some of these effects can be reduced by performing the assay in low pH (3) or with high detergent and salt concentrations. Some reports advise the test to be performed in pH (1.5). In such cases, the standard curve should also be prepared in similar matrices. Moreover, absorbance should be recorded immediately as otherwise the DMMB-GAG complex starts precipitating.

10. RNA or ssDNA can bind to Quant-iT™ PicoGreen® reagent, which can be removed by treating the sample with DNase-free RNase.

Acknowledgments

This work has been supported by National Science Foundation Award # 1624515. Pallab Datta acknowledges the Department of Science and Technology, Government of India, INSPIRE Faculty Award. The author thanks Dr. Weijie Peng (Penn State University) for his fruitful insights and Dr. Adil Akkouch (The University of Iowa) for his assistance with SEM imaging. The authors also thank Dr. James A. Martin (The University of Iowa) for providing facilities with mechanical testing.

References

1. Dababneh AB, Ozbolat IT (2014) Bioprinting technology: a current state-of-the-art review. J Manuf Sci Eng 136:61016. https://doi.org/10.1115/1.4028512

2. Ozbolat IT (2015) Scaffold-based or scaffold-free bioprinting: competing or complementing approaches? J Nanotechnol Eng Med 6:24701. https://doi.org/10.1115/1.4030414

3. Wu Y, Wong YS, Fuh JYH (2017) Degradation behaviors of geometric cues and mechanical properties in a 3D scaffold for tendon repair. J Biomed Mater Res A 105:1138–1149. https://doi.org/10.1002/jbm.a.35966

4. Yu Y, Moncal KK, Li J et al (2016) Three-dimensional bioprinting using self-assembling scalable scaffold-free "tissue strands" as a new bioink. Sci Rep 6:28714

5. Ozbolat IT, Hospodiuk M (2016) Current advances and future perspectives in extrusion-based bioprinting. Biomaterials 76:321–343. https://doi.org/10.1016/j.biomaterials.2015.10.076

6. Ozbolat IT, Moncal KK, Gudapati H (2017) Evaluation of bioprinter technologies. Addit Manuf 13:179–200. https://doi.org/10.1016/j.addma.2016.10.003

7. Gudapati H, Dey M, Ozbolat I (2016) A comprehensive review on droplet-based bioprinting: Past, present and future. Biomaterials 102:20–42. https://doi.org/10.1016/j.biomaterials.2016.06.012

8. Murphy SV, Atala A (2014) 3D bioprinting of tissues and organs. Nat Biotechnol 32:773–785

9. Peng W, Unutmaz D, Ozbolat IT (2016) Bioprinting towards physiologically relevant tissue models for pharmaceutics. Trends Biotechnol 34:722–732. https://doi.org/10.1016/j.tibtech.2016.05.013

10. Nicholson JK, Connelly J, Lindon JC, Holmes E (2002) Metabonomics: a platform for studying drug toxicity and gene function. Nat Rev Drug Discov 1:153–161. https://doi.org/10.1038/nrd728

11. Yu Y, Zheng H, Buckwalter JA, Martin JA (2014) Single cell sorting identifies progenitor cell population from full thickness bovine articular cartilage. Osteoarthritis Cartilage 22:1318–1326. https://doi.org/10.1002/nbm.3369

INDEX

Alberto Rainer and Lorenzo Moroni (eds.), *Computer-Aided Tissue Engineering: Methods and Protocols*,
Methods in Molecular Biology, vol. 2147, https://doi.org/10.1007/978-1-0716-0611-7,
© Springer Science+Business Media, LLC, part of Springer Nature 2021

Printed in the United States
by Baker & Taylor Publisher Services